KEVIN CHEN

4D PRINTING

TRANSFORMING THE FUTURE BUSINESS ECOLOGY

RC
Books Beyond Boundaries
ROYAL COLLINS

4D Printing: Transforming the Future Business Ecology

Kevin Chen

First published in 2021 by Royal Collins Publishing Group Inc.
Groupe Publication Royal Collins Inc.
BKM Royalcollins Publishers Private Limited

Headquarters: 550-555 boul. René-Lévesque O Montréal (Québec) H2Z1B1 Canada
India office: 805 Hemkunt House, 8th Floor, Rajendra Place, New Delhi 110 008

ISBN: 978-1-4878-0488-6

To find out more about our publications, please visit www.royalcollins.com

4D PRINTING

www.royalcollins.com

FOREWORD

Compared to the prevailing 3D printing, the concept of 4D printing, which has been around for more than two years, seems a bit deserted. What is 4D? Where did the additional "D" come from?

Before you start reading the following text to explore 4D printing, let me tell you a short story.

One day, a man named Adam and a woman named Eve stole the forbidden fruit of the tree of knowledge, of good and evil, in the Garden of Eden. Although they didn't understand design, together they created a work which I think is the greatest in this world – the human!

In the entire process, the sperm of a man and the egg of a woman matched and combined to form a brand-new cell, and the inside's raw materials that make up the human were the 23 pairs of chromosomes – 46 chromosomes in total. Each chromosome carried a certain number of design factors that we call "genes." Genes support the basic structure and performance of human life, store all the information of race, blood type, pregnancy, growth, and the apoptosis process of each person.

After the materials were fully prepared, the first person was "printed," and the carrier of this printing was the mother of all humans – Eve. Since then, with the extension of the time dimension and "catalysis" of nature, humans had begun to change through birth, growth, disease, aging, and death. They had also started to further deduce the reproduction of lives and the important physiological processes, such as cell division and protein synthesis. All these changes and processes repeating in cycles may not be new to us and we know them well, for sure.

So, now, that we are finished with the storytelling, let's get back to business. What is 4D printing?

The above short story implies a common understanding of 4D printing, that is, the genetic code contained in each of our chromosomes is the most primitive design programming code of all of us, humans. The growing process of our life is the organizational change generated by humans, the 4D printed object, based on the fourth dimension of time and the "catalysis" of nature.

In other words, the biggest difference between 4D printing and 3D printing is that 3D printing, like all manufacturing processes nowadays, is the end of creation. However, 4D printing is the beginning of creation, just like a human birth.

Of course, this is just a small analogy to 4D printing based on personal understanding. To unveil it and give you a full and more comprehensive picture, let's take a look at how the versatile Baidu[1] explains this science-fiction phenomenon. According to the information of Baidu encyclopedia entries, 4D printing has one more "D" than 3D printing which is the dimension of time. People can set the model and time through software and the transformable material will self-transform into the desired shape over the specified time. To be precise, 4D printing is a kind of material that can transform on its own, in which the design is directly embedded. Without connecting any complicated electromechanical equipment, it can self-fold into the corresponding shape, in line with the product design.

Is this really how 4D printing works? Let's take a look at its past and present.

FOREWORD

On February 26, 2013, at the TED2013 conference in Long Beach, California, Skylar Tibbits, a lecturer from the Department of Architecture, Massachusetts Institute of Technology (MIT), used a novel combination of materials on a 3D printer to create a string of strand. When the strand was placed in the water, its shape changed and formed the letters "MIT." This is a technology that produces the effect by combining water-absorbing polymeric materials with alkaline plastics and is defined as 4D printing. "What we're saying here is, you design something, you print it, it evolves. It's like naturally embedding smartness into the materials," Tibbits said in an interview.

On October 8, 2014, the U.S. Foreign Affairs bimonthly magazine published an article entitled Preparing for the 4-D Printing Revolution. The author, Nayef Al-Rodhan, is an honorary fellow at St. Antony's College, Oxford University, said in the article that the possible applications of 4D printing were endless, and the real future of digital manufacturing lay in the fourth dimension, which was to print objects in programmable materials that could self-transform over time.

Based on this, we may soon see that science fiction could turn into reality.

Imagine that the underground of every city we live in is full of piping systems, the fixed capacity, and high maintenance costs have always been troubling the constructers of every city, too. However, with the arrival of 4D printing, by using programmable materials, each pipeline will be able to adapt to the changing environment, and its capacity and flow rate can be adjusted through expanding or contracting motion. It can even self-repair when damaged or self-decompose when scrapped.

More amazing is that the application of 4D printing in the field of biomedicine will greatly change the existing medical conditions. In the era of 3D printing, we have been able to use 3D printed hearts, limb skeletons, and other human organs, to replace the aging and diseased organs of the human body to keep humans healthy. With the emergence of 4D printing, it will undoubtedly bring us more possible optimized solutions. For example, with the change of a heart stent, no surgery is needed if we can achieve a 4D printed liquid stent and inject it into the programmed smart material via the blood circulation system. When it reaches the designated part of the heart, it will self-assemble into a new heart stent. Moreover, the latest scientific research has announced worldwide that humans' DNA can be modified. This fact leaves aside the moral arguments of using and modifying DNA because at least a window is opened for humans to maintain health. While for 4D printing, this will be like a golden key to open this window.

CONTENTS

CONTENTS

CONTENTS

1

THE CONCEPT

On February 25, 2013, at the TED2013 conference in California, USA, Skylar Tibbits from MIT showed the world a gorgeous transformation with a string of strand — 4D printing technology. A concept that was about to change the world was born.

Skylar Tibbits demonstrated the self-transformation process of a 4D printed object in the aqueous medium through a complete experiment and elaborated its principles in detail. The 4D printing technology, which inherits the footsteps of 3D printing, no matter from the perspective of future technological development, or the perspectives of mining the value of current technologies and exploring the future business direction, is more foresightful and disruptive than 3D printing for human development. Not only is it a revolution of production tools, but also a technology derived from the change of production materials that will change the business' ecological structure, the model of medical science development, and the state of humans' lives within the entire society of the future. Therefore, it is not just a disruption in manufacturing technologies but also the future of the entire humankind.

Section 1 | 4D Printing is Here

4D printing technology was first introduced at the "Technology, Entertainment, Design (TED)" conference held in Los Angeles. Skylar Tibbits, a scientist from Self-Assembly Lab at MIT, formally brought it to us by demonstrating the self-transformation and self-assembly of a set of products, which made us aware of this new technology. A common understanding of 4D printing is that the time element is added based on the fundamentals of a 3D printer.

In the traditional concept of dimensions, it is difficult for us to cognize and comprehend the concept of the fourth dimension. From 1-dimension, 2-dimension to 3-dimensions, these are the dimensions based on physical space and can be intuitively seen with our sensory vision (Figure 1-1). However, the arrival of the fourth dimension seems to transcend our current understanding of the three-dimensional space. Its newly added dimension, time, is what we have never seen or touched but a concept that always

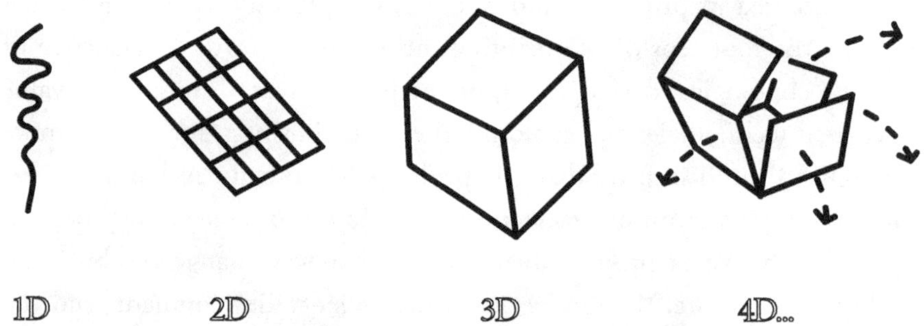

Figure 1-1 From 1-dimension to 4-dimension

surrounds us. This is one of the explanations of 4D printing now — 4D printing is printing time.

This is not an illusory concept of the future, but a new technology demonstrated at the TED conference that is being researched by MIT. Throughout the demonstration, a 4D printed strand in multi-material completed self-transformation and self-assembly on its own while being dipped in water. As introduced at the conference, the printer used to print this multi-material is not a new magical technology but just a 3D printer. However, the input of the raw materials is the magical part which is composed of a plastic and a layer of water-absorbing "smart" material. Tibbits thinks that "the printing process is not something new and the core value is the change occurred after printing." And that is exactly the key to 4D printing.

Skylar Tibbits, the director of the 4D printing project, is only 28 years old and is an architect and a computer scientist. The 4D printing method he invented can only work in the water at this time, which means that the triggering medium is limited to water only. According to Tibbits, self-assembly is the core behind 4D printing and it has been used on the nanoscale for many years, just like the self-folding proteins that are now widely applied (Figure 1-2). If the applications of the technology are further extended to the construction of bridges, tall buildings, biological organs, and daily necessities, this would turn our perceptions upside down. The extension implies that the applications of 4D printing are not far away from our daily life but they are close enough to touch, and soon they will comprehensively reach us.

The concept of 4D printing is derived from biology. The 4D printing technology demonstrated by Tibbits is a set of self-assembling particles designed with 3D printing and embedded magnets under the guidance of Dr. Arthur Olson, a professor of biomolecular science. Just shake the flask vigorously and the particles inside will self-assemble into a 3D model of poliovirus as if they are conscious. This experiment is the prototype of 4D printing, which also proves the feasibility of the technology.

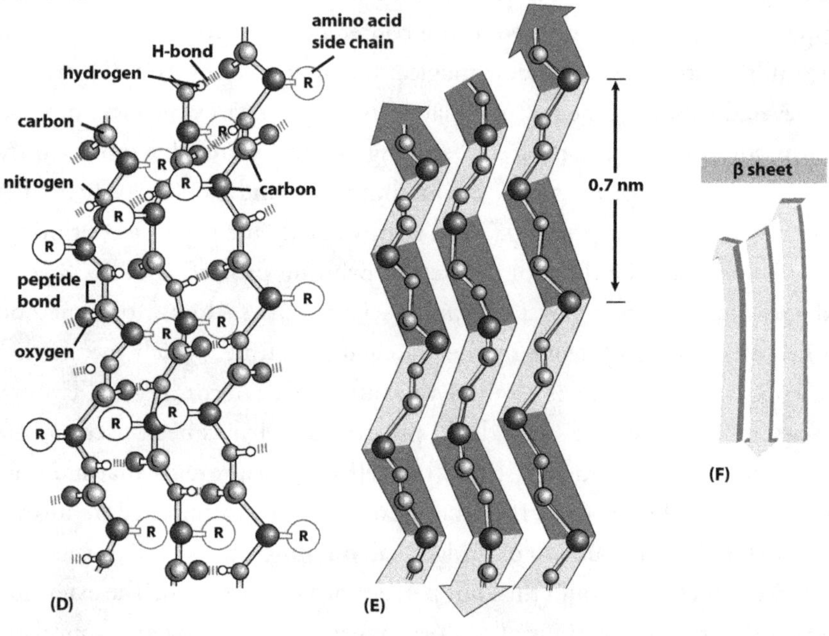

Figure 1-2 Self-folding proteins

From that point on, Tibbits had begun to explore the commercial applications of 4D printing and visited the project manager of the world's largest 3D printer manufacturer – Stratasys. In their communication, he got to know that the company's research in the field of printing materials had made a breakthrough. A macromolecule polymer that could self-transform in the water was invented. The unique feature of the new material is that it can self-extend to twice its length when exposed to water.

It is this new material that escalated Tibbits' research and development of the new technology which is 4D printing. Along with the design software and 3D printer, he mixed the two materials of extensible and non-extensible and printed into a strand, then immersed it in the water. Immediately, the strand underwent a fascinating process of self-bending and self-transforming on its own and formed the abbreviation of Massachusetts Institute of Technology "MIT" (Figure 1-3) as if it had the self-thinking ability.

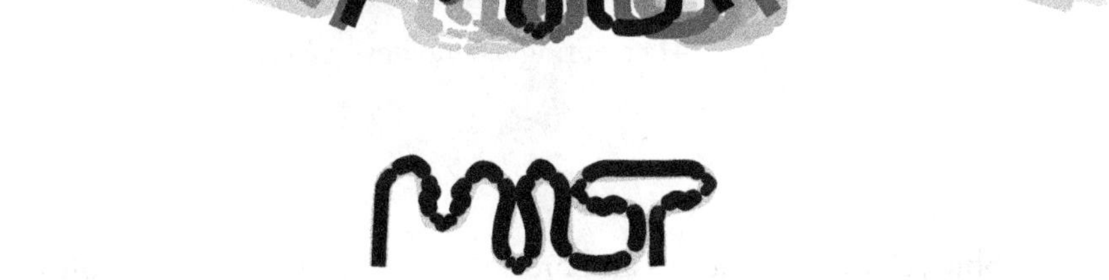

Figure 1-3 4D printed "MIT"

On this basis, Tibbits conducted the second experiment which was more complex and had more complicated algorithms and design. If the first exploring experiment was on the plane level, this time it was at the level of three-dimensional space, that is, the self-assembly and self-transformation of objects in three-dimensional space (Figure 1-4). The second experiment was also about printing the transformable design into a strand and placing it in the water. With the help of water as the triggering medium, it managed to self-assemble and self-transform into a cube.

Figure 1-4 4D printing technology of MIT

To Tibbits, significant success and breakthroughs have been accomplished in the second experiment, which marked the arrival of 4D printing. This technology will soon walk out of the lab and come straight to our daily life. The changes brought by 4D printing will be more profound and thorough. This will include the troubling underground drainage system in the cities.

If the technology is applied to build contractible drainage pipelines, the pipelines will automatically become larger to better facilitate drainage when a typhoon or a rainstorm comes. Then, the pipelines will automatically contract to their original size (Figure 1-5) when the water flow slows down.

Figure 1-5 Self-transformable pipelines

These pipelines can also self-bend, self-twist, and self-transform in specific locations in accordance with the conditions of the underground. We do not have to worry if they will burst underground, nor do we have to be troubled by the flood discharge capacity when floods come because they will self-transform by conforming to the amount of water flow. Furthermore, in areas where geological disasters occur frequently, such pipelines can even self-assemble and self-repair on their own.

Section 2 | 4D Technology vs 3D Printing

Compared with 4D printing, 3D printing is no longer a new concept, as it has been around for a while. In the early days, most of its applications were coupled with the processing center on designing prototypes to provide services for research and development and industrial design creation. However, there were drawbacks to materials, performance, and accuracy. Other than the limitation of industrial technology development, 3D printing was undeniably affected by the limitations of capital, talents, and society's understanding of the technology. With the change of the entire society and business forms brought by the mobile internet, including the rise of Makers,[2] they have contributed to the explosion of the 3D printing industry, most notably so in 2013.

Of course, the outbreak of 3D printing is the trend of the industry's evolution, and it being able to be recognized and accepted quickly by the public is related to our "dreams." Most of us might have heard of the fairy tale *The Magical Paintbrush – Story of Ma Liang*[3] when we were kids, and we must yearn to have such a magic paintbrush so that we could draw our dreams and make them come true. Although this is just a fairy tale, when combined with 3D printing, we seem to see the moment that it could turn into reality. We can print any item that we want via 3D printing, no matter if it is human, organ, food, or weapon, as long as we can design a three-dimensional model on a computer then a 3D printer will print it out for us.

It doesn't matter if we don't know how to use a three-dimensional design on a computer. We only need to scan the entity of the item that we want to print with a 3D scanner and a 3D model will be automatically generated through the computer system, then send it to a 3D printer for printing. The progress of technology has always been faster than the pace of our understanding. When we are still at the cognitive level of a 3D printer, it has quietly developed to the desktop level and is now moving towards the direction of household use. Moreover, smaller 3D printing devices have been presented to us, such as a 3D printing pen, which is just like Ma Liang's magic paintbrush whereby the process of drawing is the process of printing.

Therefore, we can have a good understanding that 3D printing is like Ma Liang's magic paintbrush, which can not only print models but also ideas for us. With 3D printing, we can design a special cup based on our ideas and creativity, and turn it into reality. We can also create a unique pair of shoes and privately customize it. We can even print a table of edible and artistic-shaped food. As we can see, the development of 3D printing has brought great influence, changes, and imagination to our future life and business forms.

If 3D printing is given the value of a dream because it is a magical paintbrush, 4D printing is the Monkey King's Jingu Bang,[4] which must be an artifact that everyone is familiar with and longing for, though, it only exists in the novel. But today, the Jingu Bang that once existed in the novel has walked into our real-life with the arrival of 4D printing. The only difference is that the new technology is 3D printed and has the function of Jingu Bang.

In the world of 4D printing, the printing object can automatically transform with a specific time set or under specific conditions, or else, it can be self-transformable and it can even self-assemble. What we print could be a cylinder bar, however, it will self-transform into a three-dimensional square when triggered, or self-transform into some other shape.

Imagine that for people who travel frequently, luggage of different sizes is required based on different trip conditions. So, we may have multiple luggage prepared at home or barely use the luggage that is not suitable for the trip. But when we print luggage through 4D printing, these troublesome situations are gone. The luggage can self-expand and self-shrink by keeping up with the pressure generated by the capacity of the belongings in it, and it can self-transform and self-assemble based on the different forms of the things loaded in it. This is the world of "Jingu Bang" technology brought by 4D printing and it will soon come into our daily life. Let's say 3D printing will affect and change business forms, then the changes and impact caused by 4D printing will be even more far-reaching and profound.

Section 3 | Digital Manufacturing

Nervous System is a generative design studio that works at the intersection of science, art, and technology. It has been using 3D printing for the Cell Cycle project since 2009.

By Jessica Rosenkrantz, co-founder of Nervous System, Cell Cycle is a web-based app that allows people to design their own, complex, honeycomb-shaped jewelry and sculptures, which can also be sold and 3D printed online.

Jessica Rosenkrantz said that the digital manufacturing process of 3D printing could produce complex and organism forms of objects that couldn't be made by other methods, additionally, it could also print unique customized products. Examples include 3D printers printing a necklace without breaking it, not to mention making whistles with peas and bottles with boats. Some 3D printers can even directly print out the operative parts, such as a multi-part gearbox.

From the introduction of Jessica Rosenkrantz, we can see that the applications of 3D printing will eventually lead to the development of "Direct Digital Manufacturing (DDM)." That is, DDM will use 3D printing to manufacture the final products or some of their parts. What's more, DDM has gained popularity in different industries, such as aerospace, jewelry manufacturing, dental technology, toy production, branded furniture design, and the fabrication of customized fashion items like sunglasses. 4D printing which follows the footsteps of 3D printing is likely to become the successor and it embeds the relevant digital parameters into the raw materials of the printed items beforehand.

The application of 4D printing is like the birth of human beings. Men's sperm and women's eggs are matched and combined to form a brand-new cell and the "genes" contained on each chromosome storing all the information of race, blood type, pregnancy, growth, and process of life apoptosis. When the living individual is born from the mother, they immediately begin to experience the changes in birth, growth, disease, age, and death under the influence of the genetic code. This will further deduce the reproduction of life and the physiological processes of cell division and protein synthesis, among other functions, that are repeating in cycles.

Section 4 | Spatial Intensification

The application of 3D printing has created great changes to existing technological innovations and industrial practices. Still, it is far from reaching the acme expected by the industrial technology revolution. At this point, what kind of expectations will 4D printing bring to human society?

The creation of a "concept model" based on 3D printing allows products to be displayed in a physical form in the early stages of manufacturing, which greatly improves the accuracy of product design and model-building. Yet, more physical printing has raised strict requirements on the printing space and skills, and it greatly constrains 3D printing to move towards "popularization." As a result, 4D printing that has developed and evolved on top of 3D printing allows product design and model-building to be simplified to the greatest extent so that physical printing is no longer confined to its formed volume and space.

A few days ago, several buildings built using 3D printing were unveiled in Suzhou, which included a villa with an area of about 1,000 square meters, a five-story residential building, and a simple exhibition hall. The walls of the buildings were printed by a large 3D printer. It only takes one day to print out a house and the results are amazing. Today, this 3D printed villa which is priced at RMB[5] 80 million, only costs over RMB 1 million.

The printer that makes the dream of printing a villa real is a three-story high "giant monster" with a height of 6.6 meters, a width of 10 meters, and a length of 32 meters, of which the foundation is as large as a basketball court. However, the width of the printed material can only be 1.2 meters. Refer to the computerized design drawings and plans, a huge nozzle controlled by the computer ejected "ink" just like it was icing a cake, and the "ink" was stacked in a "Z" shape layer by layer. Soon, a high wall was built. After that, the walls were built up like blocks and were connected by infusing with reinforced concrete in the second "printing."

The 3D printing technology makes this dreamy idea happen but the giant printer and limited-size prints will undoubtedly constrain the applications of 3D printing. At this junction, 4D printing, which possesses developed technology and extended dimensions upon 3D printing, largely solves the problems of 3D printing in mold, volume, and product irregularities incurred during product fabrication.

In the early stages of printing, which is product design, the designer has implanted the raw materials with designed parameters such as trigger medium, time, etc. By doing so, the printed product triggered by a certain medium can obtain transformation and reconfiguration to realize the shaping and restructuring of the product, which will minimize the difficulty of product printing and maximize the possibility of printing products in different spatial dimensions.

Let's build a villa in Suzhou as an example. What would it be like if 4D printing was applied? First of all, the printer is no longer the "giant monster" with big size and aggressive height but a more compact and exquisite printing technology. During the printing process, by implanting the higher-tech design parameters, it allows the printed products to have a transformable ability without the constraint of the 1.2-meter width. The printer can also print out more compact products which may extend infinitely when triggered by the medium. Just like Monkey King's Jingu Bang, with the medium of a spell, "big" or "small," it can become as small as a silver needle or as large as the sky. The 4D printed products can not only change in size but also change in shape.

Within a limited space, infinite possibilities are printed out with the use of the design and implantation of different parameters. All these products that have been designed and implanted with various changing parameters in their raw materials will bloom and display in the required time and space dimensions, completing their perfect journey of 4D printing. All the enchanting tech and incredibleness of 4D printing is simple but not easy.

Section 5 | The Overlay Scan

If we want to actualize 3D printing, in addition to printers and materials, we must have 3D models. However, as far as the current situation is concerned, there are still some professional problems in the design of 3D models, and it is difficult for general consumers to produce decent 3D models.

A company called Volumental has in-depth ideas in this regard, mainly aimed at allowing general consumers to complete their desired design of 3D models by use of a software. The app provided by the company lets people make 3D models simply and swiftly. A consumer just needs to take a photo of the object that he or she wants to print in 3D with a depth camera, and then transmit the data to the app, and eventually, obtain the 3D model on the app.

The commercial application of the 3D scanning technology provided by Volumental is very widely implemented, such as product customization. Through this process, consumers can pass the 3D data of their feet to the merchant, and with the given data, the merchant will be able to customize their shoes accurately. Similarly, consumers can also scan the 3D data of their entire body to have custom fitting clothes, or even with their stereo pictures of the same proportions.

Volumental and Intel jointly announced a plan to work together to reform the industry of selling shoes online. The concept is to sell shoes based on the 3D data of consumers' feet, and this is only the first step of their collaboration. In the future, they will further collaborate to create more new business opportunities. Meanwhile, more and more companies have begun to apply the 3D data of consumers' bodies to their commercial field.

Along with the bigger extension space and more changes generated by 4D printing, there will be more space for the software that auto-produces design depending on the scanning technology. For instance, a consumer can take two different photos they like, one of the initial model and the other of the assembled model, then the software will auto-design and synthesize the assembling process with the aid of scanning technology.

In regards to clothing, in the near future, not only will we be able to customize clothes, but also simultaneously integrate the dressing codes of different social occasions into one piece of clothing through 4D printing. To meet the different needs of social occasions, a consumer can trigger the predetermined instructions of the worn clothes with the respective medium device at any time to let it complete self-transformation and self-assembly in a short while. Such change won't be constrained by time, and for women with a desire for good appearance the applications of 4D printing on things like jewelry, shoes, etc., will make their lives more colorful.

Section 6 | 4D in Experts' Eyes

There was an article titled, *Preparing for the 4-D Printing Revolution,* published on the website of U.S. bimonthly magazine *Foreign Affairs* on October 8th, in which the author, Nayef Al-Rodhan, is an honorary fellow at St. Antony's College, Oxford University. It was stated in the article that the possible applications of 4D printing are endless.

According to the summary of the article by the Aviation Industry Development Research Center of China, the following aspects were most prominent:

1. 4D printing is a combination of 3D printing and novel material, which is even more revolutionary.

The nature of 4D printing is to print materials that can change over time by the use of 3D printing. The latest report of the Atlantic Council proposed that 4D printing not only had the economic, environmental, political, and strategic significance of 3D printing but it could also "map" the digital information (programming information) of the virtual world to specific objects in the physical world. The article on the *Foreign Affairs* website pointed out that research has shown the real future of digital manufacturing lies in printing objects that can change over time in line with the programming information.

The 3D printed objects are static and require manual control; while the 4D printed objects can automatically respond to the environment and self-assemble, self-repair, or self-transform on their own. The progress in this area will depend more on the material itself than on printing technology.

2. 4D PRINTING HAS BEEN PRELIMINARY LAB-VERIFIED AND HAS APPLICATION PROSPECTS IN MANY FIELDS.

Massachusetts Institute of Technology has preliminary lab-verified the simple 4D printing technology. Skylar Tibbits, an architect from the institute, combined water-absorbing polymer materials with alkaline plastics to create a novel material that can be used with 3D printing devices, and a string of strand was produced by printing the materials on a 3D printer. When the strand was placed in the water, it self-transformed into the letters "MIT" (the abbreviation of the "Massachusetts Institute of Technology").

The article on the *Foreign Affairs* website believes that 4D printing integrates various techniques, methods, and disciplines and has broad application prospects. Suppose that 4D printed products can be used in the underground piping system which has been plagued by the factors of fixed capacity and high maintenance costs for a long time. Using the programmable materials, each pipeline can adapt to the changing environment, while its capacity and flow rate can be adjusted by self-expanding or self-extracting, even self-repair when damaged or self-decomposition when scrapped.

Besides, the article especially pointed out that 4D printing could be used in national defense. The U.S. Army has begun to utilize 3D printing to make new equipment for the frontline in Afghanistan, though, 4D printing can allow the military to manufacture automobiles of which the coating could change its structure to adapt to any terrain. It could also help to make a soldier's uniform that could protect against poisonous gas.

3. 4D PRINTING MAY CAUSE SOME APPLICATION RISKS AND NEGATIVE EFFECTS.

As mentioned in the article on the *Foreign Affairs* website, like many other emerging technologies, 4D printing also has worrying risks and negative effects, especially the potential risks that may arise in its application in the biological field. For example, researchers at Harvard University are already using 4D printing to create nanorobots with DNA strands to fight cancer. Because the tools used in this application can be easily obtained in today's world, some people may use this technology to create new biological weapons. Besides, many risks associated with the wide implementation of 3D printing applications will also be accompanied by the development of 4D printing, such as being used for criminal means (pistols and handcuff keys have been previously produced through 3D printing). In the end, even though 4D printing can provide industrial manufacturers with more ways to customize products so that the supply chain can be further shortened, it will also endanger and complicate the technical work, product accountability, and intellectual property questions.

We have unlimited imagination about the infinite possibilities of 4D printing applications. Suppose that an underground piping system that is bothered by the fixed capacity and high maintenance costs over the long term can be fixed by using the programmable materials. Each pipeline can adapt to the changing environment and adjust capacity and flow rate by self-expanding or self-contracting. The pipeline can even self-repair when damaged or self-decompose when scrapped.

Let's suppose the possibility of applying 4D printing to the national defense industry. The U.S. Army has begun to utilize 3D printing to make new equipment for the front line in Afghanistan, and 4D printing can allow the military to manufacture automobiles with the coating that could change its structure to adapt to any terrain, or a soldier's uniform that could protect against poisonous gas.

Of course, 4D printing will not exist without danger and negative effects. The concerns about 3D printing still exist now and there is a possibility that the technology can be taken advantage of by criminals (pistols and handcuff keys have been produced by 3D printing in the past). Manufacturers will have more ways to customize products to further shorten the supply chain, which may endanger the technical work and make product accountability and intellectual property issues more complicated.

Like many other emerging technologies, 4D printing integrates various techniques, methods, and disciplines. The most worrying risks are most likely to appear in the biological field. Researchers created nanorobots with DNA strands to fight cancer by applying the principle of 4D printing. In this regard, their dual-use also brings real concerns. Due to the easy access to the essential tools, some people may use this technology to create new biological weapons.

2

THE MATERIAL

In the progress of 4D printing's development, the pace of scientific research is constantly moving forward with fruitful achievements. New memory materials from abroad and liquid metal from China combined together to fill and enrich the material-base of 4D printing. At the same time, technology research and development that depend on material technology development are also stepping forward. When our designers are still building computer models for new products to render the graphics and refinement to the greatest extent, a 3D model that allows us to have a close look at it all around gives humans greater confidence to maximize their product design. This is because when we are looking at a design on a computer, many hidden defects are not so obvious. However, when the design displays them through a physical conceptual model, the designer can truly see and touch all around the product to make sure that all of its defects are completely exposed.

Section 1 | This is 4D Printing

The key to 4D printing is not the printer but it is about having the design embedded into the printed model through software. Thus, design software has undoubtedly become another key technology after the printing materials. As we know, the disruption brought by 4D printing is far-reaching, to seize the opportunity and fully apply and promote this new technology, the R&D team of Autodesk Software Company specially designed a new software called Cyborg. By applying the principles of computer simulation technology, self-assembly and programmable materials, a simultaneous design is generated to help users realize the design optimization and handle the material folding.

The magic of 4D printing is very different from 3D printing. The innovations of 3D printing are the printing technologies and materials, while the innovations of 4D printing are the model design or model programming and materials. But, the materials used for the two printings are not currently at the same level. The logic of 3D printing is to pre-model, then print out the model through a 3D printer, which is completely different from 4D printing. For 4D printing, its steps are to embed the design and time it into the transformable smart materials via 3D printing, and triggered by a specific medium, the printed model or material should then proceed to self-assemble.

Throughout the entire process, it does not require humans' assistance or intervention to actualize its assembly, and there is no need to implant electronic components or electromechanical equipment while printing, either. It is via a specific kind of pure material triggered that it conducts self-transformation and self-assembly to eventually generate the initial model set by the user. In other words, the 4D printed primary model could be a piece of board (Figure 2-1) but triggered by a specific medium, it will self-assemble into a table or a chair over the specified time (Figure 2-2). During the process, no human labor or any external tools are needed, and it completely depends on the self-assembly of the material which carries the designs embedded in advance. This is precisely the charm of 4D printing whereby it is able to disrupt the traditional manufacturing and commerce industries, as well as the service industry.

Figure 2-1 Plastic Board

Figure 2-2 Plastic Chair & Table

Section 2 | Implanting Wisdom into Materials

Under the traditional thinking and manufacturing process, when the status, suitability, and functionality of different parts of the product are required to be inspected during the design process, a functional prototype of the product needs to be produced by a skilled technician with the use of labor-intensive workshop techniques. Commonly, the mass production of such prototypes takes several weeks and thousands of dollars or euros. In contrast, with the application of 3D printing, it only takes a few hours to produce the required functional prototype. Therefore, product design and production become fast and convenient, and their costs are then reduced to the greatest extent as well.

When we are still immersed in the benefits of changes brought about by 3D printing, 4D printing will make a fundamental change in rapid modeling. Compared with 3D printing which steps include pre-modeling, scanning, followed by printing into a physical model, 4D printing directly embeds the design into the material that simplifies the creation process from "conceptual design" to "physical object." It will allow objects to be created "automatically" like Transformers[6] and after being printed out, the desired products will self-create without connecting any complicated electromechanical equipment.

Just like all objects exist over time, the shape-shifting objects take time, too. Under the dimension of time, the shape of an object processed with special materials can change on its own while being triggered by some external stimulus. Different from the process of 3D printing which is pre-modeling and printing with materials, the logic of 4D printing firstly prints out a rigid smart material on a 3D printer, then combines the material with the above-mentioned external stimulus to complete its shape change, according to how it was designed earlier.

Traditionally, the creation process is generally about simulating first and then manufacturing, or adjusting the simulation effect while building. But, through the close integration of hardware and software, 4D printing has disrupted the traditional way of creation. 3D printing not only makes creation simple, fast, and accurate but also compresses the creation cost ten times, or even one hundred times. In terms of creation method, 4D printing is an upgraded version of 3D printing which makes the traditional creation process invisible.

The core of "4D printing" is to allow printed objects to change in shape through self-adjustment. As a result, printing will no longer be the end of the creation process but a turning point instead. "What we're saying here is, you design something you print it, it evolves," said Skylar Tibbits, director of the research team for the 4D printing project at Massachusetts Institute of Technology. He said, "It's like naturally embedding smartness into the materials." A car is a solution to rainy days and a coffee cup can also adjust itself according to the different temperatures of coffee. Just as Tibbits shared with the TED audience, "Imagine if water pipes could expand or contract to change capacity or change flow rate, or maybe even undulate to move the water themselves." 4D printing will realize the possibility of objects transforming into infinite kinds of shapes and specifications under certain trigger mediums.

Section 3 | Making the Products "Wayward"

4D printing is unlike the current popular robots and artificial intelligence. We may have worries about artificial intelligence that is in the future, whether people control machines or machines control people. But for 4D printing, such worries are quite unnecessary as the shape-changing ability it gives to the objects is only specified to self-assembly, which does not affect the creation of new things that have self-transformation wisdom and adaption to changing environments. Additionally, it will thoroughly change the current 3D printing and its related industries, such as industries of construction, furniture, piping, clothing, toys, military, etc. Hence, compared with the current hot 3D printing, 4D printing will have broader development prospects and influences.

A good example is IKEA furniture (Figure 2-3), which focuses on its design and convenient assembly, which has obvious advantages in transportation and assembly compared with traditional furniture, yet, we still need to spend labor on assembling it. However, the situation is completely different when it comes to 4D printing. 4D printing is not only more intelligent but it can print objects that have the ability of "self-creation." What we buy maybe is just a piece of the multi-layer board but when shipped to the designated location and within the specified time, by being exposed to the trigger medium which could be water, gas, sound, etc., it will be triggered to self-assemble into a shape (Figure 2-4) without human labor or external force.

Figure 2-3 Ikea Furniture

Figure 2-4 Self-assembling Furniture

Skylar Tibbits believes that 4D printing can, based on 3D printing, give the printed objects the ability to change their shape as per different environmental factors (such as sound, light, heat, water, etc.). Skylar said that the key to 4D printing is the materials that are already known and can change in response to different environmental factors, like memory alloys. By combining these materials with how we want the object to change, we can finally print it out.

The ultimate goal of 4D printing is to realize the science fiction scene discussed above. If we buy a set of furniture at IKEA, we don't need to assemble it, and it will self-assemble by itself.

Section 4 | New Printing Dimensions

While we are still discussing how amazing 3D printing is, and how it will bring changes to traditional manufacturing, the fields of 3D applications have been expanded constantly from architectural models, such as product models, simulated organs, aerospace components to delicious food, etc. Also, it has changed industrial printing to desktop printing. While everyone is talking about how 3D printing has been affecting our daily life, a more amazing printing technology, 4D printing, is already on the way to us. Other than the three-dimensional spaces of 3D printing which are length, width, and height, there is one more dimension newly added to 4D printing – "time." That means 4D printed objects no longer only exist in a fixed form, instead, they can change shape themselves according to the time set and with a help of a certain trigger.

The most intuitive understanding of 4D printing that we are familiar with today is the demonstration of the concept at the TED conference earlier by the initiator of 4D printing, Skylar Tibbits, a lecturer from the Department of Architecture at MIT. When two 4D printed strands were placed in the water, an incredible change took place. Of course, if we only observed from the surface of the printed strand, there was nothing unique and it was just a very ordinary printed material. However, the amazing change occurred after it was placed in the water, that is, it slowly curled and transformed itself, and then self-assembled into the word "MIT" (Massachusetts Institute of Technology).

Another experiment was also with a strand. When placed in the water, it seemed to come to life and stand up. Moreover, it continued to evolve and fold in the water, eventually, it self-assembled into a cube on its own. Of course, such transformation has been preset by printing the desired shape into the primary model and triggered by the external medium, which in this case was water. In the end, it managed to self-assemble into its desired shape accordingly. It's worth noting that as there is no human labor involved in the entire transformation process, it looks like the material has self-awareness, and this is the enchantment of 4D printing.

Even so, the more important key to achieving 4D self-assembly is not the printing itself but the materials. The materials used in the MIT 4D printing experiments are two different types. One is a high-molecular polymer that can extend to twice its original length in the water, while the other can have a fixed shape in the water. Once the materials contact water which is the trigger medium that has been embedded in with the time set, the expanding part of the materials will be triggered to self-transform and together with the fixed part, they will self-assemble into the designed shape. Of course, the trigger medium is not limited to water.

Therefore, we can intuitively understand the enchantment of 4D printing, that is, its use of a self-transformable material based on 3D printing. Users can complete the modeling through the software and input the attributes that they want into the material, and with a trigger medium, the transformable material will complete self-assembly in line with the pre-settings. Perhaps in the future, the trigger medium does not even have to come from external factors but could trigger inside of materials via printing.

The significance brought about by 4D printing has both, impact on the manufacturing and commercial ecological chains. It will pre-set the result of targeted assembly or molding in the primary model, then the printed object could be triggered by external stimuli within the defined period to perform self-assembly. In the future, it will not only be self-assembled but will also evolve into a form similar to self-fabrication. Or, we may also understand this in a way that 4D printing could implant a kind of memory wisdom into the material and with the impact of the medium, memory could be triggered to realize the object's assembly.

In this way, maybe one day, we can see a scenario whereby the 4D printed objects could be like robots with no wires or motors to help humans complete projects which are dangerous or beyond human's strength. This could include the installation of equipment in outer space or deep-sea, or the construction of skyscrapers. The way that the 4D printed objects differ from robots is that the objects are not electricity-driven or controlled by the back-end chip programming. Instead, they depend on the models pre-designed with parameters.

The current 4D printing is still in its infancy. With the rapid development of technology, it may not take long for many scenes in science fiction movies to truly happen in our daily life through the utilization of 4D printing. Just like how we react to the smart wearables, the arrival of 4D printing may come straight to us overnight when we are unprepared. What 4D printing is going to change is not only the way of assembly but also self-adaptiveness. In the future, the items we use will become super items, like a coffee cup that can adjust according to the temperature of coffee, a sofa that can change its size in line with the different sizes of guests, a car that is able to self-transform into an aircraft when driving into a lake to help us out of danger.

Section 5 | New Design Software

If 3D printing is one of the eight technologies that will affect the future, 4D printing is one of the technologies that affects future business. With a 3D printer, any image can be printed into reality. While with 4D printing, our ideas will be even more imaginative. The development of 3D printing will inevitably facilitate the exploration and application of 4D printing, including the promotion of design software.

As the world's leading 3D printing company, Stratasys has merged Objet and Makerbot, two of the world's leading 3D printing companies, who are the leaders in the industrial and consumer 3D printing markets. The acquisitions are going to establish the position of Stratasys in the entire 3D printing market. Further, according to 3D Print news, Stratasys later announced that it would acquire the world-famous 3D design community website, GrabCAD, in cash, and so far the two parties had reached relevant agreements with the transaction amount up to 100 million US dollars. After the acquisition, the co-founder of GrabCAD, Hardi Meybaum, will continue to manage the business of GradCAD.

GrabCAD founded in 2010, is one of the world's leading 3D design community websites, which mainly provides designers and engineers with the files and shared services of CAD models to help teams in their collaboration of project development. More than one million engineers are currently using the service, and more than 520,000 CAD models are available for free.

New opportunities will be created in the field of 4D printing which is how it will differ from the services of CAD models provided by GrabCAD. What 4D printing needs are the self-transformable factors merged with time, but not the current single static 3D model.

In the process of popularization and application of 4D printing, if you want to let 4D printing into the life of the public, it is necessary to eliminate the professionalism of design. That is to say, any user should be able to input their ideas into the model design platform, and the design software should be able to produce designs per different self-assembling schemes.

Let's say we want to print a wardrobe which is multi-layered in its original state (Figure 2-5). When placed in a designated position and given a trigger medium, the multi-layers will drive, change, and assemble into the designed wardrobe (Figure 2-6) on its own. Other than the material itself, the self-assembling model has become the core of the entire process. The model

Figure 2-5 Multi-layer

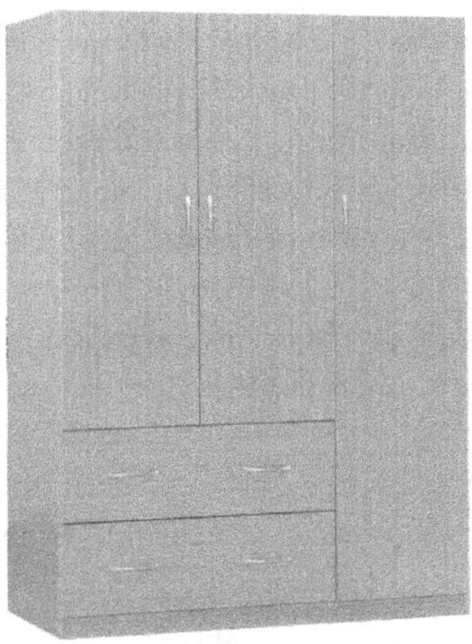

Figure 2-6 Wardrobe

connects two sections, of which one is the initial end and the other is the assembled end, and the role of driving and connecting in between is then the actual model design process. During the process, the route and result of self-assembly are embedded and triggered by a certain medium.

The model design needs to be implemented with the help of computer-aided software. Hence, the development of 4D printing will lead to the further development of CAD software, which is no longer limited by the current static three-dimensional data. Instead, it is drive by the model design with variable factors that are merged with time. Maybe the process needs to incorporate artificial intelligence technology as well.

Section 6 | Self-Transformable Composite Material

It is only a matter of time for the technical problems in the physical world to be solved, and the key lies in the accurate presentation of concepts and their sharing. 4D printing is an important direction for the development of the future. Despite this, it is yet to enter the commercial field and it is still at the exploration stage, so the pace of its development cannot be stopped. The research team led by Jerry Qi, associate professor from Department of Mechanical Engineering, University of Colorado, Boulder,[7] and Martin Dunn from Singapore University of Technology and Design, mixed polymer fibers with the capability of "shape memory" into the multi-material used in traditional 3D printing to fabricate another kind of composite material that can transform into various shapes, just like the scenes in the movie *Transformers*.

According to Martin Dunn, in this experiment, the initial configuration was created through 3D printing, and the printed material was given the design wisdom while printing. Then, the programming activity in the "shape memory fiber" began, which meant that the given design wisdom of the material started to work, and the fourth aspect of the material was generated – the self-assembly.

The concept of 4D printing was firstly proposed by Skyler Tibbits, founder of the Self-Assembly Lab at MIT, in April 2013. According to his explanation, 4D printing is about "self-assembly," that is, the material automatically turns into a predetermined model. His research team once put a strand made of multi-material into the water, and it transformed into its pre-designed shape. This is the basic experiment of 4D printing we are aware of and the one that proposed this idea.

Martin Du from the University of Colorado said, "We advanced this concept by creating composite materials that can morph into several different, complicated shapes based on a different physical mechanism. The secret of using shape memory polymer fibers to generate desired shape changes of the composite material is how the architecture of the fibers is designed, including their location, orientation and other factors."

The popular understanding is that based on a material with "shape memory" functions, or maybe a material with intelligent logic, its transformable structure could be designed through software which includes the joints and connections of the transformation. After that, the 4D printed object could be triggered to self-transform under a specific medium.

One of its core elements is material. Currently, the materials meeting the requirements of 4D printing can be described as scarce. If the materials for 3D printing are still looking for a solution, for 4D printing, they are still at the start of their search. But, the emergence of technology will consequentially trigger and promote the revolution and development of new materials. The materials for 4D printing will create huge opportunities in the future, of which the commercial value is no less than the value of the "gold mine" big data.

Section 7 | Microstereolithography

In the development of 4D printing, the results of scientific research are constantly advancing. Recently, researchers from the Massachusetts Institute of Technology (MIT) and Singapore University of Technology and Design (SUTD) have developed new 4D printing that allows printed objects to "automatically" change from one shape to another within a few seconds. All it takes is a little change to the temperature of the environment.

Many researchers have been trying to use 3D printers to fabricate shape memory objects, but because the fabrication process remains at the millimeter level, it limits the speed of material recovery. Nowadays, when the "Microstereolithography" technique with high-resolution projection is applied, the design of objects can be carried out on the micron level (1 mm = 1000 microns, which is equivalent to the diameter of a hair), which makes the recovery speed of the material extremely fast. In short, the more the material is made on the micro-level, the faster it can be recovered, with the recovery speed as fast as a few seconds. For example, a wildflower can release all its pollen in a few milliseconds because its trigger mechanism is measured in micrometers.

At the same time, researchers have also been looking for an ideal polymer combination to make a shape memory material adaptive to the lithographic pattern. They selected two polymers, one shaped like a curved pasta, and the other crisscrossed like a scaffold on a construction site. They mixed the two polymers, and the resulting new material could withstand huge stretching and twisting without damage. The characteristic of this new composite material is that it can withstand huge deformation and can be stretched up to three times the original shape, which is even greater than any existing printing materials.

The new material can "bounce back" to its original shape, that is, its shape when it just came out of the printer. When it is exposed to a temperature of 40 to 180 degrees Celsius, it can be restored to its original shape within seconds. If we could reduce the entire fabrication process to a smaller level, we might be able to shorten the recovery time to a few milliseconds.

Section 8 | Liquid Metal from China

Presently, the Cryogenic Biology and Medical Lab at Technical Institute of Physics and Chemistry reported for the first time the free shaping effect of liquid metal. It could be stably presented in any shape on the surface of graphite, and it could achieve a climbing movement in reverse gravity.

Previously, due to the large surface tension of the metal droplets, they usually existed in a spherical shape in the electrolyte, and their shaping ability and transformation mode was relatively limited. Although the liquid metal can temporarily change its shape by applying an electric field, once the external field is removed, the liquid metal will rapidly retract into a spherical shape under the action of its surface tension – unable to maintain the previous structure. Now, scientific research has found that by introducing a graphite substrate, the liquid metal in the electrolyte can be flexibly shaped into various sharp patterns, such as strips, triangles, squares, rings, and more arbitrary shapes. The free spreading and shaping effects of liquid metal will provide new directions and more possibilities for the development of soft deformable robots and the application of 4D printing to a wider range of fields.

Experiments have also shown that when a drop of liquid metal is placed on the surface of graphite immersed in a sodium hydroxide solution, the liquid metal will automatically spread out to form a flat pancake shape. This is very different from the liquid metal presenting in a spherical manner on a glass substrate. The reason for this spreading effect mainly comes from the electrochemical interaction between the liquid metal and the graphite substrate. In alkaline solutions, the surface of graphite is usually positively charged, and the surface of liquid metal is negatively charged. When these two conductors come into contact, the charge will flow from the liquid metal to graphite. The surface of the liquid metal is then oxidized to form an oxide film, which significantly reduces its surface tension, and the liquid metal now behaves like mud, which can be shaped into various shapes at will. This realizes the free shaping of liquid metal in an open liquid environment. It breaks through the original liquid metal component control mode and holds an important value in the design and even advanced fabrication of unshaped flexible electronic devices, transformable intelligent machines.

Section 9 | The Self-folding Device

4D printed objects can self-transform in the dimension of time through the action of a certain external medium, and the printing materials are responsible for this. Shape Memory Polymers (SMPs) can realize the memory of a certain shape. After printing, under normal conditions, the object will maintain a fixed state. When evenly heated, it will change to another planned shape. By printing a variety of materials with different dynamic mechanical properties into a 3D object according to customized drawings, you can create an object that can change its shape in a controlled, predetermined order.

Through 4D printing, the reaction of a smart shape memory material to temperature can be preset in its programming to achieve the effect of the self-folding of complex structures. That is, when the components are heated, each SMP would change shape at a different rate, depending on its internal clock. Accurately designing the chronological sequence of these materials to change shape allows three-dimensional objects to assemble themselves.

Figure 2-7 shows the self-folding process of smart shape-memory materials, which responded slightly differently to temperature. This is done using materials with slightly different folding rates to ensure that the components do not interfere with each other when folded. The components can respond to stimuli such as temperature, humidity or light in an accurate and timed manner, thereby forming a three-dimensional structure, deployable medical equipment, robot toys and other various structures.

"Previous efforts to create sequential shape changing components involved placing multiple heaters at specific regions in a component and then controlling the on-and-off time of individual heaters," explained Jerry Qi,

Figure 2-7 Smart Shape Memory Materials

professor from Mechanical Engineering at George W. Woodruff[8] (George W. Woodruff School of Mechanical Engineering). "This earlier approach essentially requires controlling the heat applied throughout the component in both space and time and is complicated. We turned this approach around and used a spatially uniform temperature which is easier to apply and then exploited the ability of different materials to internally control their rate of shape change through their molecular design."

An important aspect of self-folding is the control of the self-collision of objects. Care should be taken to prevent the different parts of the folding structure from contacting and hindering subsequent folding. In this regard, existing research has created a metric to predict collisions, which is used together with the simplified model to design self-folding structures. Such structures can be stably locked into the desired shape. For example, an unmanned aerial vehicle can be in shape when cruising and can be transformed into another designed shape when diving. Alternatively, the 3D device can self-fold into a flat shape or self-roll into a tube for easy transportation, and then self-reassemble into a predetermined 3D structure when used.

3

THE PRODUCT

Bright flowers, with lasting bloom – 4D printed. Healthy people, young and energetic – also 4D printed. Such examples are beyond enumerating. The 4D printed gadgets are not just about satisfying humans' daily necessities, but about making the impossible, possible. By unveiling 4D printing technology, the world thus becomes more colorful.

Section 1 | Feast on the Table

One of the reports that has received much attention earlier was the artificial eggs made by a Silicon Valley start-up called Hampton Creek. The eggs were sold out after being marketed in Hong Kong. The reason why this company was concerned, other than the artificial egg itself, was the investor of the company, Ka-shing Li.

The artificial eggs and ordinary eggs are not very different in taste, but their cost is reduced by 48%. The important thing is that they do not contain cholesterol, so they are nutritional, without unhealthy ingredients (Figure 3-1).

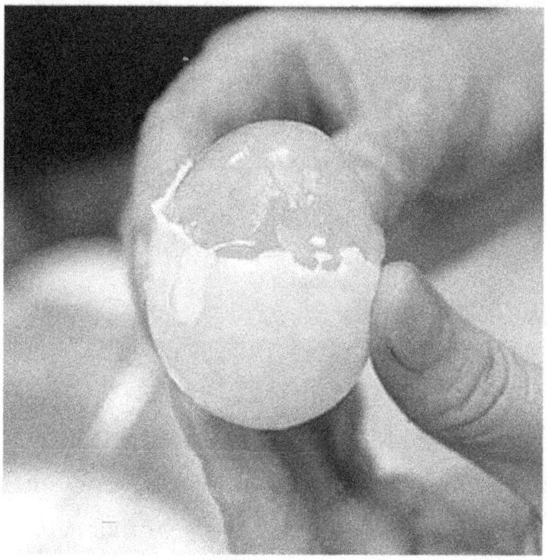

Figure 3-1 Artificial Egg

Figure 3-2 Artificial Beef

Today, Ka-shing Li, the company's main investor, has invested $10 million in another company called Modern Meadow, which is working on making artificial meat through 3D printing. The difference is that Modern Meadow uses 3D printing technology to cultivate artificial meat in the lab. A mixture of different types of cells constitutes a "biological ink," which is then molded using 3D printing and finally processed in a biological reaction device to produce artificial meat (Figure 3-2).

In addition to Ka-shing Li, the legendary Silicon Valley billionaire, Thiel, is also one of the investors. He invested in US $160,000 to US $220,000 in the venture through Breakout Labs, a subsidiary of the Tel Fund.

1. How big is the market for artificial meat?

According to the figures compiled by Modern Meadow, people consume about 300 million tons of meat food annually worldwide, and by 2050, this figure is expected to exceed 500 million tons. In the United States, the average person consumes about 120 kilograms of meat per year, which is equivalent to 1,000 burgers.

Artificial meat can be used for household consumption, on the other hand, it can be printed directly specified to different hamburgers, without secondary processing as is done for traditional beef. For other edible food, such as snacks, it will also be more convenient. The advantage is that it retains the elements needed by the human body and eliminates excessively high-fat elements.

2. Printed artificial meat is just around the corner.

In fact, artificial meat is not a technological concept. However, this new technological thing may change the meat production industry and could be part of an important means to solve the global food crisis. Andras Forgacs, CEO of Modern Meadow, said that the cells obtained by the company come from small living tissue and do not harm or kill animals. From the currently announced artificial technology, generally, the following steps are taken:

1) Biopsy, collection of skin and muscle cells as samples;
2) Process pancreatin, destroy cell adhesion, disperse tissue into single cells, and expand culture;
3) Integrate the cells together in sheets, and glue them together in the subsequent biopolymerization process;
4) Place the integrated cell layer in an in-vitro reaction vessel and continue to mature;
5) After a few weeks, terminate the nutrition supply of the cells and collect the cells to process to finished product.

For Modern Meadow, the techniques applied in the entire process are not too difficult. The first two steps involving the in-vitro cell culture technology are already mature, and can even be implemented by commissioning to relevant professional companies. Among all, the most important steps are the third and fourth steps. The third step involves a technology called 3D Bioprinting, which is not a problem in terms of the current development of technology.

Therefore, compared with genetically modified foods, artificial beef may be a better way to solve food shortages. The most important difference between artificial meat made by 4D printing and by 3D printing is that by 4D printing, it brings more convenience into our daily life. Imagine that we will no longer need to slice meat, and we will just need to preset "self-separate" in the initial model, and when needed, just put it in the water and wash it, and then chunky beef will be automatically be separated into slices (Figure 3-3).

The replacement of 4D printed artificial meat makes different changes according to the different needs of customers, so that the dining table is no longer a taste feast, but also a visual feast.

Figure 3-3 Self-transformable Beef

Section 2 | Leather Bags

It is easier to apply the printing technology to leather than meat products. Mainly from a technical point of view, the structure of the skin is simpler than muscles and therefore easier to produce.

From the biological point of view, the organization of meat is much more complicated than the manufacturing of fur. The fur is closer to 2D, but production of meat requires a more complex 3D technology. In addition to technology, the supervision of cultivating leather is relatively simple. A lot of products can be designed after artificial leather is produced, such as belts, leather bags, leather clothes, among others. With the help of 4D printing, the texture of the leather will self-transform according to the different social needs and occasions for customers. Therefore, the decorative leather products will be full of variety.

In the near future, women would not have to prepare too many bags for different social occasions. All they will need is a 4D printed bag that may change into many varieties to meet different social occasions. Besides, compared to the widespread application of artificial meat, the main reason that 4D printing of leather will be easier lies in its relatively loose regulatory conditions.

Section 3 | The Revolution of Underwear

There are significant developmental changes during the different stages of female breast tissues. Even in adulthood, when breast development has matured, many women have breast fullness, stiffness, and tenderness before menstruation. In severe cases, the slightest movement can be very painful and uncomfortable. This is due to the increased estrogen in the body before menstruation which leads to breast hyperplasia and edema.

The current female underwear is similar to traditional clothing and shoes. Its size is fixed, and will not change according to the changes in breasts. Some researchers used uniform 3D printing to print underwear (Figure 3-4). The

Figure 3-4 3D Printed underwear

unusual aspect of the bikini is that it is printed by a 3D printer. The outline of the underwear is "woven" by circular patches connected to each other, and the thinnest part is only 0.7 mm. By using nylon 12, its texture is very soft and not easy to break. According to the manufacturer, this underwear is very refreshing and comfortable to wear after it is soaked, and it may be more suitable for swimming.

As the 4D printed underwear can self-transform and self-assemble, it will change the current the fact that the size and shape of the underwear cannot adapt to the changes in breasts. Comfortable underwear-wearing experience will be available through the 4D printed underwear that can self-transform according to the changes in breasts before and after menstruation. This may completely change the entire market of female underwear.

Section 4 | Fashion and Jewelry

The fashion industry has a long history, especially in womenswear, of which there's never been a shortage in the consumer market – it comes in bulk, customized, with various sizes and styles. Yet, the only problem that cannot be solved is making clothes that fit all curves.

The short history of 4D printing technology does not affect its development. The U.S. Massachusetts technological design studio Nervous System recently developed an elastic fit fabric and printed it into the world's first "4D dress" by using 4D printing technology (Figure 3-5).

The dress solves the problem of ill-fitting clothing, as it changes itself according to the wearer's body shape, and, more amazingly, it changes its shape automatically (Figure 3-6). This is perhaps the dress that is missing from every woman's wardrobe. The fabric fibers used to make the dress are made of 2,279 triangles linked by 3,316 built-in hinges. The tension between the triangles and the hinges varies with the shape of the human body, so no matter if you gain weight or lose it, this 4D dress will never fail to fit.

Figure 3-5 4D Printed Dress

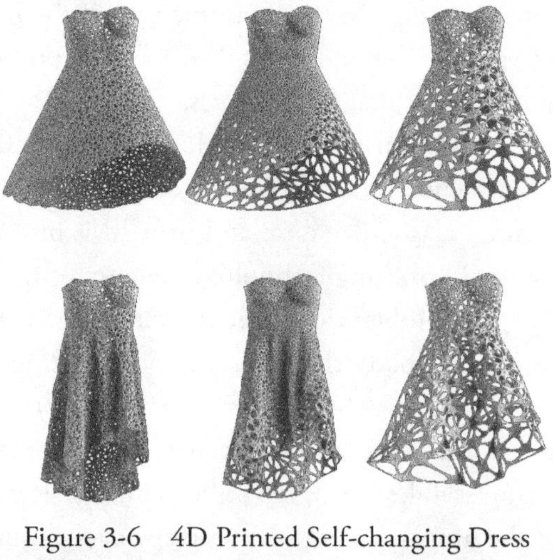

Figure 3-6 4D Printed Self-changing Dress

The space taken up by the dress inside the printer is much bigger than printing a regular 3D model dress, but technicians still managed to print it out in one piece. Meanwhile, researchers also 4D printed a series of jewelry to match the dress, which undoubtedly could also change automatically based on the shape of the human body.

The principle of 4D printing material used by Nervous System was a breakthrough based on MIT's model. It applies "Selective Laser Sintering Technology" that uses a laser to sinter powdered nylon material to create prototypes. The powdered materials between the triangles and the linked hinges are not sintered by the laser and drop after printing, and thus, interlocking fibers are created. It takes 48 hours to produce each dress, of which the cost is GBP1,900 (about RMB18,393). However, the price does not discourage women.

The 4D printed dress means that the shape of a 3D printed object can transform automatically into another shape, allowing the object to transform into its final design form without manual labor. Nervous System has developed an application that enables users to 3D scan their body, and then select a fabric size and shape, and instantly customize a unique 4D dress.

A long time ago, we were able to print parts through 3D printing and then manually put them together to create a large object. However, what makes 4D printing different is that the printed object can be automatically assembled or transformed into a predetermined shape. In contrast to MIT's idea, they named this process "Kinematics" after the branch of mechanics of the same name — also known as the geometry of motion, which describes the movement of an object rather than its cause.

Rosenkrantz said, "We think the greatest advantage of 'Kinematics' is that it can transform any three-dimensional shape into a flexible structure for 3D printing. The system then compresses the structure down through computational folding."

To make these dresses, a 3D scan of a person's body form is essential and is the basis of the digitally modeled garment. The hardness and the state of the final product can be controlled at this stage by selecting the structure of the triangular hinged mesh, and the way the material drapes can also be simulated on-screen. With computer simulation software, this digital model can be folded into a much smaller shape and then printed out in a compressed form. When the dress is lifted out of the printer, it will automatically unfurl into its predetermined shape.

According to Rosenkrantz, "Compressed designs offer benefits not only for production but also for transport. It holds great promise for the creation of flexible wearables but could also be used to enable the production of other large-scale structures in today's small-scale printers." Last year, in response to the request from Motorola, Nervous System has begun developing the concept of Kinematics to customize 4D printed products.

Kinematics can generate a design composed of 10 to 1,000 different parts which can be expressed to one another to build a dynamic or mechanical structure. When printing a dress, for example, first, one must scan a user in 3D to sketch the dress, and then inlay a pattern on the dress, followed by generating the Kinematics structure to simulate the draping features of the dress. Finally, the dress which is much bigger than the printer, is compressed and then computationally folded. The dress can be printed in one piece instead of different parts and combined into one.

The price of these products varies depending on the customization options. As long as customers are happy with their customized design, after placing an order, Nervous System will print it out. The second free application allows users to experiment with the Nervous System's templates and print the results at home. Rosenkrantz and Louis-Rosenberg further developed their theory by adding the ability to fold the design into its smallest spatial structure.

The advantages of 4D printing are, on the one hand, the ability to compress designs into their smallest spatial structures and then 3D printing them, so that the printed objects exclude redundancy. On the other hand, the printed objects can change to fit around different needs. All these benefits are unmatched by 3D printing.

The current operating pattern of the fashion field will be changed through the implementation of 4D printing. For example, no clothes have to be displayed, and no inventory is needed. Instead, a human body 3D scanner, a computer with professional design templates, and a virtual reality mirror will be needed. Users could then choose the template from the computer-based choices on their style and preferences, and the template could be of any style or even their photo. Then, the system will automatically design and generate the corresponding garment.

The model auto-generated precisely by the system with the help of a human 3D scanner will be scanning the size of the human body and then merge it with the design. Customers will then be able to see a simulated display that is close to the real. Through the virtual reality mirror, customers could rotate the model for display at any angle and adjust and modify it as they wish. Finally, the garment that they had in mind would be generated. After that, the virtual garment will be printed out by transmitting it to the printer via the cloud service. Alternatively, if the store has a 4D printer, the garment could be printed out directly in the store. As such, 4D printing will bring fashion into the era of private customization.

Section 5 | Sex Toys

If both 3D and 4D printing want to enter the sex toy market, the following problems need to be addressed.

First is the problem of safety and health. For example, among the current 3D printed products, especially those with low-end molding, they face low-precision problems and are often very rough on the surface. In the field of sex products, users will hardly be able to get satisfaction with such products. Besides, the rough texture can easily breed bacteria. These problems can't be ignored in 4D printing.

Second is the problem of privacy. In terms of the current climate, it takes quite some time for 3D or 4D printing devices to enter a household. Users still need a relevant service provider to process their personalized sex products, which involves questions about the user's privacy. This may become a considerable concern while users are selecting their service provider.

The third is the problem of materials. The materials used for sex toys are mainly silicone, particularly used for the parts that have direct contact with sexual organs. The current materials used for either 3D or 4D printing are unable to satisfy the actual usability of sex products, especially the key technology, "smart" materials, of 4D printing. But, as technology advances, the problem of materials will be solved.

As such, if 4D printing cannot reach the household application level, even in Europe, America, and Japan where sexual concepts are relatively open, people will not be willing to discuss sex products in public and will also be embarrassed about receiving them. After all, it is a private matter.

On the other hand, just as there are no two identical leaves in the world, each person's physiological structure is also different. So, the sex products already in the market may not be suitable for all, which is why 3D printing is currently popular.

The emergence of 4D printing technology makes it easier to privately customize sex products. The products can be designed and printed based on the individual's physiological structure. In addition, a 4D printed sex product can self-change while being used along with the user's physiological response during the different stages to facilitate an optimal experience.

When the design technology is further developed, a human body model can be generated and designed with only a photo. The most significant difference between the 4D printed human body model and the inflatable doll is that the model can be designed and printed with the image of a user's dream lover. Meanwhile, it would also self-change with the different physiological responses of the user during use. No matter what cup size you want to experience, A cup, B cup, or D cup, the 4D printed sex toy can self-change to bring you comfort and pleasure.

Similarly, for female users, regardless of what kind of smart "egg," smart "stick," or 3D customized toy is used, they are hardly able to make adaptive changes to the different physiological responses of the human body. Hence, for the sexual organs that are sensitive and delicate, it is difficult to provide the maximum "intimate" care.

The situation is entirely different with 4D printing technology. It doesn't matter if it is an "egg" or a "stick," as long as the trigger is set while printing, which can be the secretion of sexual organs, a kind of breathing rhythm, a sound or an intensity, among others, this magical item will self-change its length or size with the changes of user's physiological response.

If we add a 4D printed sex product to the robot, in the future, do we still need real partners? Without a doubt, when 4D printing is implemented intensively in the field of sex products, a new ethical discussion will be inevitable.

Section 6 | Fashionable High Heels

It is not news that 3D printing technology has made its way into the fashion industry. A collection of 3D-printed women's high heel shoes (Figure 3-7) was exhibited at a futuristic fashion show at the World Fashion Center in

Figure 3-7 3D-Printed High Heel Shoes

Amsterdam. It comes from a graduation project of Pauline Van Dongen, who is a graduate of a Master's in fashion at ArtEZ University of the Arts in Arnhem, Netherlands.

As seen from the below figure, there is one thing that is common with 3D printed shoes – they are overall seamless. The mesh design breaks the traditional impression of shoes and feels nondurable. However, that's not the case. The shoes are actually printed from nylon, which is not only lightweight but is also strong and tough. What's more, their leather insole is patented, and the entire outer layer of the shoes is coated with a synthetic rubber, which can increase the friction, improve the durability, as well as the comfort of wear.

At the same time, people in related industries from China are actively involved in not only printing villas, but fashionable shoes, as well. The entrepreneurs in Tianjin printed fashionable shoes by using 3D printing technology (Figure 3-8).

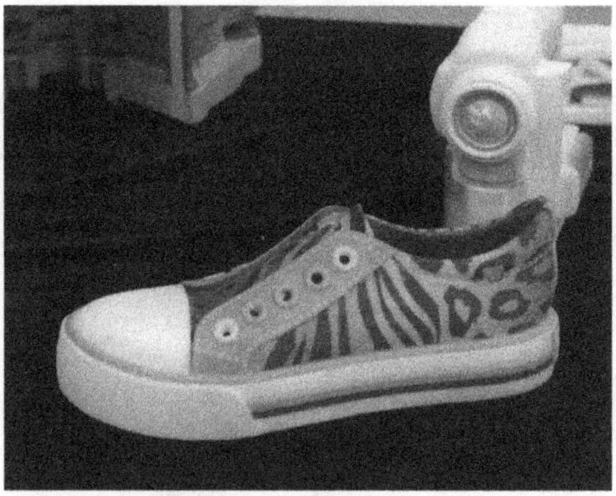

Figure 3-8 3D Printed Shoes in China

Neither 3D printed shoes nor traditional shoes can solve one problem, that is, the independent variable of shoes. In other words, the shape and size of the shoes we buy are fixed, and will not change according to the thickness of the socks we wear, nor will they change their shape and size while we are climbing or walking.

With 4D printing, however, this is a different story. 4D printed shoes will thoroughly change our definition and perception of shoes.

A pair of 4D printed shoes (Figure 3-9) from MIT's self-assembly lab, formed entirely on their own, and could even contract around the feet once they are put on. According to Skylar Tibbits, the leader of 4D printing in the U.S., 4D printed shoes are self-forming and self-contractible, and the contractible materials used can minimize the cost of making shoes. Additionally, the materials used in 4D printing are even more "magical" as they can change their shape in response to external stimuli, such as temperature, water, pressure, and so on.

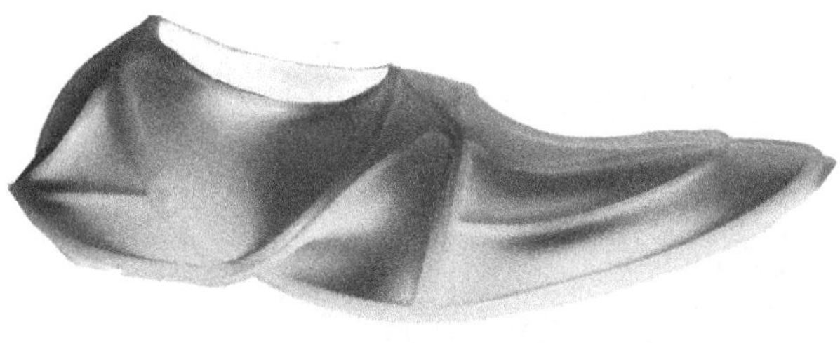

Figure 3-9 4D Printed Shoes

With such "capricious" shoes, in the future, we don't have to worry about not being able to get the correct shoes. Regardless of size or style, the future 4D printed shoes can fulfill our customization needs. After users try on their favorite shoes in the smart mirror, the 4D printer will automatically print out the corresponding shoes according to the users' foot size and density of bones. This will solve all "fitting" problems.

The shoes of the future will self-adjust their size and shape according to not only the swelling of our feet and the thickness of the socks we wear but also the different wearing occasions. At least some changes are foreseeable. First, women will no longer have to be embarrassed when their boss or friend suggests mountain climbing. The high heel shoes on their feet will automatically change into flat shoes that are suitable for mountain climbing or long-distance walking. Second, we don't have to worry about the different sizes of shoes required in different environments, nor do we have to worry about the looseness of shoes long-term wear because 4D printed shoes will automatically change and adjust to maintain the appropriate size for us. Even in the further future, the full intelligence of 4D printed shoes will be realized in its applications, like switching to ankle protection mode while playing basketball; switching to grass walking mode while walking on the grass; switching to waterproof mode while walking in the rain.

Problems like shoes' size, fit, and so on, are only the first hurdle solved by intelligent technology, 4D printing, in the field of footwear and apparel. In the future development, with 4D printing technology, more applications will bring human beings magical use and enjoyment. Nothing is impossible!

Section 7 | Self-Transformation of Toys

With the continuous improvement and enhancement of material life, the market demand for the toy industry is continuously expanding, especially the modern toy industry. Modern toys in China started in the mid-to-late 1980s when the industry emerged accompanied by the Reform and Opening Up. The toy industry is labor-intensive in the sector of manufacturing, it is the same as the industries of textile, clothing, shoes, plastic products. The so-called educational toys nowadays, also fall under the category of traditional manufacturing.

According to some of the information disclosed by the industry so far, from January to December 2013, the primary business revenue of the national-scale toy manufacturing industry was 166.44 billion Yuan, a YOY[9] growth of 14.3%; total profit of 8.33 billion Yuan, a YOY growth of 27.0%; total taxes of 4.38 billion Yuan, a YOY growth of 22.7%. China has become a significant production country of toys.

Presently, about 90% of China's toy industry relies on exports, and more than 70% of them are incoming material processing or sample processing. In 2012, the national toy exports amounted to US 114.5 billion dollars, a YOY growth of 5.8%. Although this looks good as data, from the perspective of the overall independent design and creativity of the toy industry, China's toy industry is still in its infancy. It's not unusual for a country which is mainly at the stage of industrial manufacturing to ignore the cultural and creative elements of the toy industry.

Influenced by the factors like the increase in national income, the pressure of economic downward trend and the national support for the industrial transformation and upgrading, as well as the cultural creative industry and Makers' activities, it is inevitable for the toy industry to develop rapidly. Such development must be closely related to 4D printing.

The toy industry is not limited to the traditional concept of children's toys. Toys should be thought-leading tools that bring people of different age groups enjoyment and help them explore their lives. This will inevitably lead to niche products and demand for creative and personalized products. Obviously, the traditional large-scale manufacturing fails to satisfy the new development of the toy industry. The small-batch, personalized, and rapid manufacturing will become the critical production technique that will achieve the rapid growth of the toy industry. The rise of 3D printing technology is, to a certain extent, an excellent solution to this problem.

The application of 4D printing technology may not only change the in-bulk selling ideas of the traditional toy industry but also bring the toy industry into the field of creativity. For instance, children's needs of toys keep changing, especially for those under the age of five whose curiosity and exploratory skills are quite strong but relatively weak in concentration. To satisfy their needs, some families may bring their children to professional children's playgrounds to fulfill their curiosity, while some families keep buying their children new toys.

However, such methods have brought trouble to parents, like the inefficient placement of toys in an expensive house, and difficulties of meeting the changing needs of children regardless of any methods. Perhaps, 3D printing technology can be a good choice, that is, each family can be equipped with a 3D printer, and every day, parents could print out toys from the printer with the materials and models purchased. This may be a good solution, but in regards to 4D printing technology, it is obviously not suitable.

Through 4D printing, when printed toys are not needed, the materials could be recycled and reprinted many times. Alternatively, printed toys can exist without a fixed form. Just like the popular toy nowadays, Transformers. The only difference is that the 4D printed Transformers could also self-assemble and self-transform rather than be assembled or disassembled by children.

This disruptive to the children's toy industry of children between 0–3 years old. The Makers are the people who benefit the most from the concrete changes brought about in the era of 4D printing technology. Makers are a group of people who share the same idea of pursuing innovations, and the toy industry is a challenging creative industry, especially so in adult toys, which have challenges in innovative thinking. For the Makers, on the one hand, there are many new market opportunities, and it is an industry worthy of challenges and exploration. On the other hand, there are new challenges to the children's toy market, like how can one design a toy that can self-transform into a duckling, a chick or a puppy? This is the innovation of the toy industry driven by 4D printing technology, which has far-reaching significance for the education of both, parents and their children (Figure 3-10).

Figure 3-10 Self-Transforming Toys

Besides, for traditional toy manufacturing companies, in the era of 4D printing, the business model will no longer be about selling toys, but instead, it will be about selling innovations or providing 4D printing services. Consumers could either purchase the existing creative toy models directly through the internet platform or put forward their innovative ideas of toys they want to make. Those would then be processed online by professionals like Makers or toy designers, and then printed out on the designed model either by the designer or by the nearest printing service provider appointed by the consumer. Of course, consumers would also be allowed to have the purchased models printed out by their own 4D printer.

To put it briefly, in the era of 4D printing technology, the future of the toy industry is no longer in selling toys, but in creativity. Of course, the sale is not about the materials, as the materials can be recycled, which not only solves the problem of energy-saving and environmental protection but also minimizes the consumption costs for consumers.

Most importantly, the 4D printed toys that can self-assemble and self-transform with given instructions, can not only satisfy children's curiosity and stimulate their creative thinking, but help parents better meet the needs of their children, as well.

For the adult toy market, 4D printing technology will pose a challenge for Makers and will undoubtedly inspire the industry's development. Our toys may take us into a toy mobile maze, which will have profound significance in our thinking and exploration. Likewise, 4D printing will also change the training industry, as well as the assessment of HR recruitment. Maybe soon, the interviewer will no longer give a paper-based test, but a ready-to-change 4D printed toy, like a mobile maze, where the candidate will need to find the variables and complete the toy within a specified period (Figure 3-11).

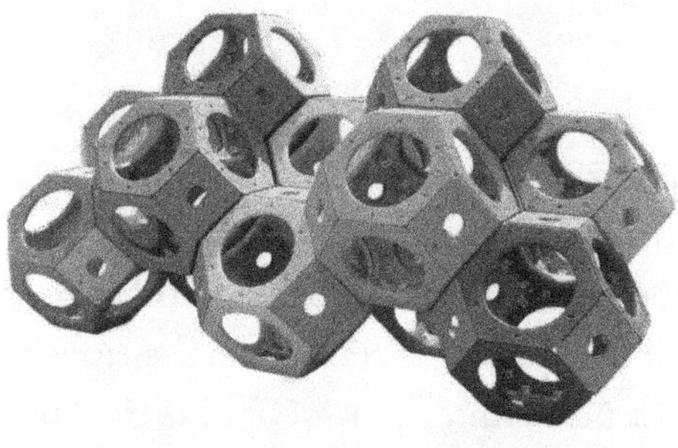

Figure 3-11 Toy with Variables

Section 8 | Disruption to Residence

The 3D printing technology has been around for quite some time, and it's only been two years since the capital boosted it and made it popular. Pushed by capital and talent, many geeks, researchers, and entrepreneurs have made various attempts based on 3D printing technology, including printing techniques, printing equipment, printing materials, and so forth. They have also explored 3D printing in different fields, such as printing of artwork with sand, printing circuits with graphite, and printing hearts or limbs. What is unique is that they used 3D printing technology to print houses, which could probably revolutionize the construction industry.

Actually, 3D printing technology in the construction field and the printing of houses is not new. Although it has indeed brought revolutionary techniques to the construction industry. Its advantages can be mainly reflected in the below aspects.

First, the most significant advantage of 3D printed buildings is energy-saving, environmental protection, and the saving of materials. This is because there is no dust or construction waste generated in 3D printing buildings, which is good news for city dwellers who are suffering from haze and construction noise.

Second, it can turn the waste into treasure and realize the secondary use of construction waste. All kinds of construction waste, industrial waste, and mine tailings can be used as raw materials for 3D printing. In particular, the sand of the desert is like "gold" for 3D printing – it is an excellent raw material. It can be used to make sand fixing walls and vertical greening walls, amongst other things, as well as realize sand fixation and help to control the desert.

Third, it can help to save construction costs and improve construction efficiency. In regards to the 3D printed buildings we have actualized so far, the application of the technology can save 30% to 60% of building materials, shorten the construction period by 50% to 70%, and save labor by 50% to 80%. That means, with 3D printing technology, it is possible to build a house at a lower cost and within a shorter period, saving at least 50% on construction costs.

In its practical applications, the concerned 3D printed buildings are in Suzhou. These include a villa with an area of about 1,100 square meters, a five-story residential building, and a simple exhibition hall, of which the walls were 3D printed using "ink" composed of a little steel, concrete, and construction waste (Figure 3-12).

Figure 3-12 3D Printed Building

1. 3D printing: How are the walls printed?

In the light of the Suzhou 3D printed buildings, the "printer" used was 6.6 meters high, 10 meters wide, and 32 meters long. Its foundation was as broad as a basketball court and it was three-stories in height. The width of the printed material can only reach 1.2 meters, but the length can be unlimited. According to the computerized design drawings and plans, a large nozzle controlled by the computer ejected "ink," similar to cream decoration, and the "ink" was stacked in a "Z" shape layer by layer. Soon, a high wall was built. Afterward, the walls were built up like blocks, then joined together by infusing with the reinforced concrete in the second "printing." In this way, ten buildings of 200 square meters each could be printed out within 24 hours.

2. Will 3D printed houses bring down house prices?

Given the current market price of an 80-million villa, the cost of 3D printing is just over 1 million. Other than the advantageous price, in terms of efficient construction, it is way beyond the traditional construction, and such a villa can be printed out in a day.

3. Are 3D printed houses safe?

Although there are no specific monitoring standards for 3D printed buildings, or relevant standards for 3D printing in the construction industry today, from the perspective of developers' tests, 3D printed walls are five times stronger than the ones printed with ordinary cement. With the help of the 3D printed walls, a more complex and solid structure can be achieved, and also the hollow wall structure can create a living environment that is warm in the winter and cool in the summer.

To obtain the large-scale commercialization of this emerging technology and its applications, the overall performance of stiffness, strength, and durability, among other factors, has yet to be further validated. Also, relevant testing standards are yet to come.

4. What will be disrupted with 4D printing?

The appearance of 3D printing in the construction industry not only enables the customization of buildings but also effectively reduces construction costs and increases construction efficiency. This will have a revolutionary impact on the existing real estate market. But when architecture meets 4D printing, the enhancements may not only be in house prices, but also the below two aspects.

First is the efficient use of living space. The space we live in today seems to be an issue that affects the quality of our lives, and all of it is due to the high house prices. To think further, the size of floor space truly has an impact, but the rational utilization of space seems more dominant. The houses we live in today, let's say the living room, are often not utilized. This includes our bedrooms. Generally speaking, while using either a bedroom or living room, the other room stands idle, whereby a contradiction arises between the inadequate space and inefficient utilization.

4D printing can efficiently solve this issue. When we are in the living room, the spatial pattern of the house could automatically change, merging the space of the bedroom into the living room to create extra room. And vice versa, when we need to rest in the bedroom, the spatial pattern will similarly auto-change, merging the space of the living room into the bedroom to provide a more comfortable resting area.

Second is the impact on the renovation industry. Renovation is not needed for 4D printed architectural space because the elements of the renovation wanted by the residents would be integrated into the initial printing and printed out directly with the printing technology. Not only that, but the 4D printed architectural style, renovation style, and renovation layout can also all change depending on the residents' ideas. We can also say that the style of home renovation can change daily with our moods.

All we will need to do is to give our ideas for changes to the model design of the building at the early stages of the printing process. Then, the environment, layout and style of the interior renovation will self-change according to the trigger instructions from the residents. In a 4D printed building, it may no longer be a building that we live in but a "magic cube."

Section 9 | Centennial BMW

Centennial BMW released a brand new concept car "BMW VISION NEXT 100" that represents the future of BMW and the direction of its latest technologies. These new technologies and materials were all produced using 4D printing. Its 4D printed body parts are functional and at the same time, by programming the printed objects, they also give the car the ability to change shape in the future.

Functionality: All of the car's 4D printed parts will be functional. Today, the design, manufacturing, and functionality of the parts are all worked on separately. While in the future, vehicles' parts will not only be manufactured using today's additive manufacturing process but they will also be completely functional after printing. We could say that the vehicle is going to be "growing" out of a pile of raw materials.

Transformability: 4D printing technology is not producing specific parts, but new materials that are smart and interconnected. The hardness and the water solubility vary depending on the arrangement of the tiny fibers within these materials. Researchers use such properties to "encode" the printed objects to enable them to transform into a more complex shape.

The innovation and application of materials will be the core of automotive manufacturing's future. Carbon fibers and new composite materials are already being used in current BMW models. The new materials used in the concept car, such as fibers made from recyclable and renewable materials, will be realized by 4D printing and rapid manufacturing of the future, while the application of these new techniques will bring earth-shattering changes to the field of automotive materials and manufacturing. Additionally, as the techniques of rapid prototyping and rapid manufacturing become more prevalent, the smart and interconnected materials produced by these techniques will make the automobile a product that is way more complex and flexible.

Section 10 | Safety Protection Equipment

An example of this is the car's airbags. In the event of a car accident, they can quickly eject to protect the people in the car, minimizing the injuries suffered by the people sitting in the front due to the impact of the car crash. However, the level of protection of the airbag is still limited, and the ejection of the airbag itself has a certain level of impact on the human body.

It would be an excellent idea then if there was an object that could be transformed at the moment of a crash to protect the human body from its impact. With 4D printing and the successful development and application of new polymeric materials, the introduction of such safety protection equipment is no longer a dream. All that's left is a matter of how long we need to wait for this dream.

The 4D printed safety protection equipment, first of all, will protect essential parts of the human body much like a bulletproof vest. Further, it will also look a lot like the legendary golden shield and iron shirt[10] (the most famous hard qigong in Chinese Kung Fu), whereby the wearer would be able to withstand punches and kicks without any damage, and even ordinary swords and knives could not hurt them. Furthermore, the wearer could also reach the levels of strong qigong to protect the entire body, so that it can obtain unimaginable effects such as not drowning in water, not burning in a fire, not dying while holding one's breath. Besides, it will be lighter, thinner, and smaller than a bulletproof vest in its volume. It can be even internalized into the human body just like how the golden shield and iron shirt, so that once it is subjected to external force, it will instantly appear to protect the human body from harm. Therefore, such safety protection equipment should not only be applied to car driving but also be worn by people in daily life just in case of emergencies.

Section II | UAV[11] Craze

Nowadays, owning an aircraft toy or a drone is nothing new. From toys to civil and military use, driven by capital, UAVs (drones) have developed just as rapidly as 3D printing. When the Chinese New Year of 2015 just started, Ziyi Zhang was proposed to by her boyfriend by placing the diamond ring on a UAV flying slowly towards her. Since then, the UAV, as a technology star, crossed over to the entertainment circle.

From the perspective of UAV development, one point of difference is that the community of Makers boosted its current craze. Up until now, we could control a small helicopter with the help of a mobile application. For people who love to stay at home, it can provide logistics service to deliver food, or by pairing it up with a camera, one could experience God's point of view by exploring nature through aerial photography (Figure 3-13). One could even propose marriage.

The current UAVs are no longer a mystery. Many of them are controlled by mobile devices, which may not even be considered new. A 3D printed UAV, driven by Makers, has taken to the skies (Figure 3-14), and the difference is that its appearance can be assembled at will with the help of 3D printing technology.

Figure 3-13 UAV Photography

Figure 3-14 3D Printed UAV

Today, except for the standard technical modules, such as the core printed circuit board for autopilot, motor, batteries, camera, among other remaining components, including the propeller and its shell, are all the results of 3D printing. Users can 3D print the shape of the shell according to their preferences to form their aircraft.

With the aid of 4D printing technology, the aircraft will bring fundamental changes that could further expand the scope of UAV applications. For example, in the event of a change in the patrol environment, a UAV can self-assemble to change its shape in response to those changes (Figure 3-15). While being in an open environment, a UAV can appear and exist in a broader form. But, when entering some narrow spaces, it can self-assemble to change the shape into either circular or superimposed, allowing exploration of some deep wells or unknown caves.

Figure 3-15 Aircraft with Different Shapes

In the face of different changes in weather conditions, 4D-printed aircraft could also undergo self-assembly and self-change, by always keeping the camera in the most rational working state to adapt to flying and performing aerial photography tasks. Of course, these functions can also be achieved by giving the aircraft electromechanical component technology. But, with 4D printing technology, the aircraft could self-drive and self-change through its material structure rather than being driven by its motor components.

Besides, the aircraft will be further impeded from moving forward in the air during its operation due to the separation of aerodynamic forces in the direction of the airflow, which is what we usually call resistance. The different parts of the aircraft will be subject to pressure drag, frictional resistance, and induced drag, caused by various factors. The pressure drag can be described as the dust that often rises behind the car while driving at high speed because the air pressure in the vortex area behind the vehicle is small, and it sucks up the dust. Frictional resistance is the air flowing against the surface of the aircraft, and due to the viscous nature of the air, it results in friction with the aircraft's surface and creates frictional resistance that impedes the aircraft from moving forward. The generation and magnitude of this resistance is related to the shape of the aircraft.

Through 4D printing technology, objects made of smart shape memory polymers will have the ability to self-fold, which will allow the unmanned aircraft to choose the optimal shape to resolve the various resistance obstacles efficiently under different flight conditions. This will create an optimal flight. For instance, it could have one shape at takeoff, one shape during the flight, and another most suitable shape when diving. This will not only minimize the air resistance to the aircraft but also reduce friction loss to ensure the best use of the aircraft.

Section 12 | Electronic Products

Along with the development and manufacturing technology of electronic products that are continually improving, represented by television sets, telephones, computers, combined speakers, and so on, all have taken the road to miniaturization. In early 2007, with the birth of the first iPhone, electronic products that once were segregated in their functions were now heading towards gradual intensification. There were more and more devices that started to integrate a variety of applications, and some even became all-around kings.

Now, with the advancement of 3D and 4D printing technologies, what kind of changes are coming to electronics?

First, new manufacturing grafted with 4D printing technology will make electronic products invisible. The new materials used in 4D printing will have flexible elements and memory functions, which will enable manufacturers to embed the programming of the functional features into the memory materials at the early stages of design, and then incorporate the triggering media for different functional features via 4D printing. This will then allow the products to present different functions of applications by triggering their respective medium.

An electronic device that needs to be idle for a while could be attached anywhere like our hair, glasses, watches, clothing, or shoes – a place that is almost undetectable by others. Once one of the functional applications is required, it can be activated by triggering the corresponding medium, which can be water, electricity, or external physical action, or it may even brain waves, so that its functions start working how they are expected. In the process, the device could further transform if needed.

Just like our current widely used smart glasses, the 4D printed lens which can attach to our retina, with the blink of an eye can start the corresponding functions, such as taking photos, video, and so on. At the same time, it can also start the thought-search controlled by the brain waves, accessing the internet to stimulate the role as an e-tourist while surfing online. It could also use the function of a VR[12] port, sending people to the ancient times that they were previously unable to experience, or the unknown future a hundred years ahead, and maybe even the distant interstellar space.

4

APPLICATIONS

Every single 4D printed product will make a little bit of difference to our future life. If you are of marriageable age, do you have the impulse to flee from the "torture" of the elders of the family always urge you to get married on every holiday? If you are a new dad, do you also have a lot of frustration and embarrassment taking care of your baby? There are many "ifs" in our life. However, when 4D printing becomes popular, it will make many of the awkward and troublesome things in your life simple and natural.

Section 1 | When Super Dad Meets 4D Printing

As a new dad, it is a joy to spend time with your baby. But when it comes to making formula, changing diapers, undressing the baby before bed, and lively telling bedtime stories… these tasks are headaches to most dads. Although the advent of a formula dispenser machine that can instantly make a bottle of warm formula has been a solution to one of these headaches, the rest are still bothering them.

1. When diapers meet 4D printing

For example, diapers, for most new dads, make the babies' upset because of their short-lived patience. It is much easier to clean up a diaper than to put on a new one. It usually takes several adjustments to put on a diaper properly and the entire process is difficult for babies who cannot focus and cooperate. If the diapers are manufactured by 4D printing technology, as long as the position of the buckle does not greatly deviate from product to product, the degree of tightness will be self-changing and self-adjusting to achieve optimal control (Figure 4-1). On the other hand, when the capacity of a diaper exceeds the normal and there is no time for a change, it will be able to self-change according to the humidity as the triggering medium. It will expand the capacity to accommodate the texcess.

2. When pajamas Meet 4D printing

Another example is that babies don't want to sleep undressed, but they can fall asleep while wearing clothes. When sleep and the changing of clothes

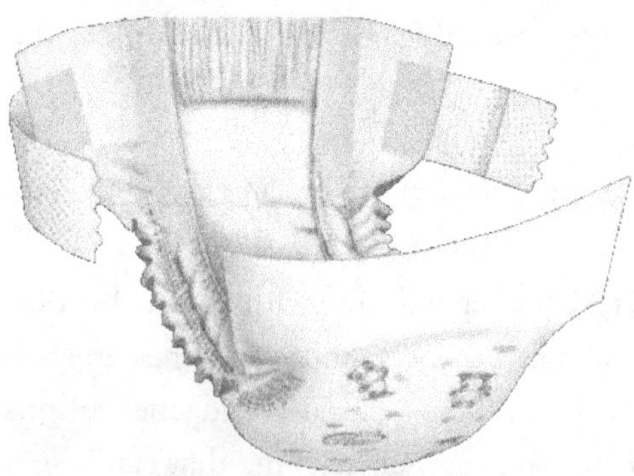

Figure 4-1 Diaper

come together, perhaps this can also trouble a father. However, if the clothes are manufactured using 4D printing technology, it could very different.

On the one hand, when a baby falls asleep, the 4D printed clothes worn by the baby will self-transform into a quilt, on the other hand, in line with the weather and the thickness of the clothes, they can self-assemble and self-transform to reach the best conditions.

Section 2 | Faxing a Boyfriend Home

Changes in technology always develop at a speed beyond our foresight. The text fax was only popular for a short while, and now the 3D printer, Zeus, is here. Now physical objects can be "faxed" as well. Then, in the future era of 4D printing technology, can we fax a boyfriend home for the Chinese New Year?

AIO Robotics has launched a 3D printer called Zeus that can scan, fax, photocopy and print 3D objects. As a standalone device, Zeus doesn't need to be connected to a computer to work. After scanning an object on the workbench of a Zeus, a digital model will be quickly generated and saved. The model can be either printed out directly from the Zeus or sent to another Zeus, just like a fax, for it to be printed in its entirety.

In other words, next time when you need a screwdriver, and if your neighbor happens to have a Zeus, you can ask your neighbor to "fax" it over to you. As the development of 3D technology is constantly maturing, and the objects that can be printed cover basic living consumption, such as clothing, food, housing, and transportation, the scenarios whereby the applications can be applied in the future will be even richer.

With the coming of 4D printing technology, what we can fax is no longer gadgets for daily necessities. As far as the 4D printer we have allows, we can fax a full set of furniture and even an entire house. What makes this faxing process different from 3D printing is that the capacity of space and the printer limit the transmission of a large 3D printed object. Meanwhile, 4D printing is different because the transmission of a large object does not rely on a self-assembled model but on a model of any minimum space, which can be "faxed" distantly to the designated 4D printer for printing and then it can self-assemble to the expected form by its trigger.

The "fax" in the era of 4D printing will exist in another form, relying on the platform of cloud services. We can fax the printing object anywhere in the world as long as a 4D printer is available.

Not only that, but it is also reported scientists in the Netherlands have successfully transmitted a subatomic to 3 meters with 100% reliability. According to scientists, there are no laws of physics that prohibit the transmission of large objects, including humans, over long distances. So in the future, it will be possible to fax a boyfriend.

Section 3 | A Home Life Package

Tools

Most people have probably been in similar situations where they were annoyed because of the lack of tools at home, such as scissors, wrenches, and other simple tools. We can print them out with the help of 4D printing technology, and what differs it from 3D printing technology (Figure 4-2) is that instead of having to make a single print for each tool, we can design the variations of multiple tools directly into the printing model. Then, we can trigger the medium to self-transform and self-assemble the printed tool.

Even the lack of a stool, chair, ladder, etc. is not a problem. Despite this fact, 3D printing technology is also sufficient for printing these objects (Figure 4-3), and 4D printing technology is incomparable in this aspect. For example, when we need a ladder, with 3D printing technology, we must print it out at a predetermined height, and when we do not need it, it will inevitably take up space. But with 4D printing technology, the ladder can be designed in an optimal and minimal form, and the height that it can extend could be set in the printing model so that the ladder could be triggered

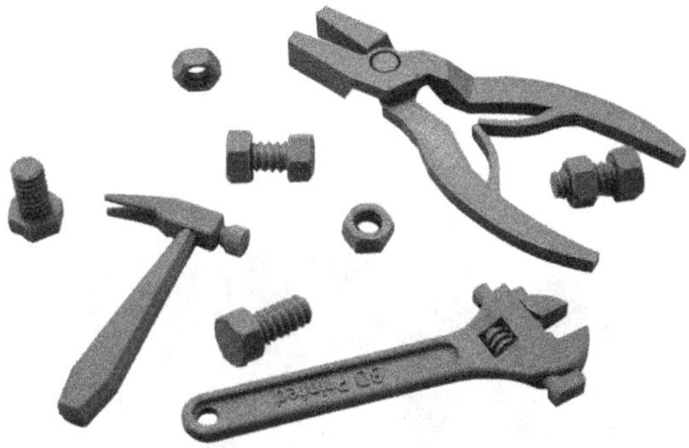

Figure 4-2 3D Printed Tools

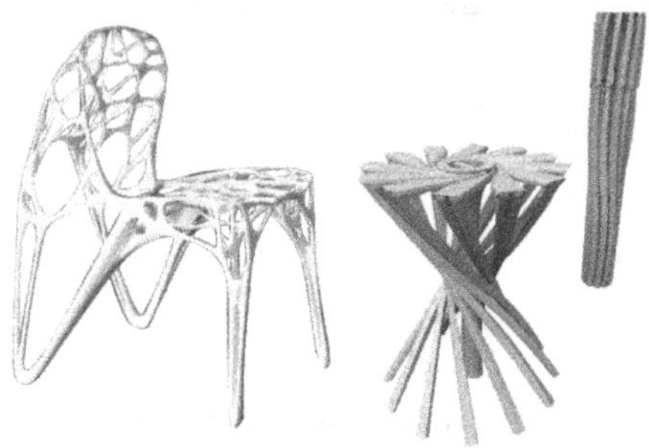

Figure 4-3 3D Printed Chair and Stool

to self-adjust to different heights to fulfill different needs under various environments. It can also return to its "original form," making it a stool or chair for everyday use when the user does not need it.

As for kitchenware such as spoons, cups, plates, pots, bowls, scoops, and others, they are not a problem for 3D printing technology (Figure 4-4), as well. However, these 4D printed objects are far beyond their 3D printed cousins. There is one significant feature of 4D printed kitchenware, that is that it can self-transform based on the amount of food cooked. We don't have to worry about the situation whereby the pot is too small for cooking sufficient food to serve our guests.

Figure 4-4 Kitchenware

Section 4 | 4D Printing for Military War

4D printing is currently focused on and applied not only in the field of people's livelihood, and the military industry has shown greater interest in it, as well. According to the U.S. Army's Chief Technology Officer, Grace Bochenek, 4D printing "takes 3D printing and adds a transformation dimension." The idea behinds this is that the properties of a 3D printed component can be altered when exposed to different environments, such as water or extreme temperatures. The applications of the idea include body armor and equipment for soldiers.

Bochenek said body armor could be 4D printed in the future. The Army has been struggling over the past decade to meet the dual needs of body armor, which is to be both protective and light enough not to burden the soldier or restrict their movement. In the future, scientists may develop lightweight and compact body armor using 4D printing that is easy to store and carry but one that can also expand and provide full protection.

Moreover, the application of 4D printing in the military field will affect the serious missions of national security. According to an article in *Science and Technology Daily on the 4D Printing Technology's Broad Prospects for Military Applications,* it will focus on the impacts of military technology and the issues of national security brought by 4D printing. The article's points can be summarized as follows.

While 3D printing technology is on the rise, 4D printing has come into our reality. In the movie *Transformers,* the scene where a car instantly transforms into a giant robot is just amazing. With the help of 4D printing technology, this science fiction may probably be realized in the near future. At the TED conference held in Los Angeles in February 2013, MIT researcher, Skylar Tibbits, placed a composite strand containing water-absorbent smart material in the water, and the strand automatically twisted and transformed into the shape of the word "MIT." As already mentioned, this was the first public demonstration of 4D printing technology and it became a sensation around the world.

4D printing technology is about the refinement of 3D printing technology. 3D printing is an additive manufacturing technology that is based on digital model files and it uses powdery metal or adhesive materials to construct an object by printing it layer by layer. Compared to 3D printing, 4D printing adds a dimension that can "change" the object, and its core technology is the self-assembly of materials. This means it can embed certain smart materials in the predetermined positions during the 3D printing process, then place the printed object in a specific environment, and physical or chemical changes will then occur under the influence of the external environment. These changes will lead to further changes in the elements of the object, such as altering its overall shape, stiffness, and so on.

Today, the broad application of 4D printing technology has attracted a lot of research in the economic fields. It also has huge potential prospects in the military fields.

The first step is to open up the link between the manufacture of weapon equipment. 4D printing technology will overturn the traditional life cycle of weapon equipment and possibly optimize it from manufacture to deployment to use, to scrap to manufacture, and back to deployment, site modeling, use, recycling, and then redeployment. At the deployment site, according to different surrounding environments and operational goals, weapon equipment will be able to enhance and adjust its design parameters and be rapidly molded. It could even self-adapt to the environments to significantly improve its environmental adaptability and operational effectiveness. For instance, through 4D printing technology, it is possible to obtain a camouflage net, which will automatically change shape and color in response to the changes in external lighting. It could better blend with the surrounding environment and improve the camouflage.

The second step is to facilitate the on-site manufacture of large equipment components. Although the prospect of 3D printing is broad, it is limited by the size of 3D printers. It can only actualize the on-site printing of components in small and medium sizes. With 4D printing, the folded form of a large structure, the essential parts, and the sensitive materials required for its unfolding can be designed in advance. Then, a semi-finished component could be printed using a 3D printer, and by the control of the special physical field, the finished component could be unfolded automatically. The typical application of this is the space required for self-assembly of large structures such as satellite solar panels, antennas, etc. This will significantly reduce the number and weight of mechanical parts needed, lowering down the volume and weight required for satellite launches.

The third step is to promote the development of micro-military robots. The microrobots will perform a large number of reconnaissance or strike missions on the battlefield in the future. The volume, weight, and energy consumption are very much related to the mechanical components required for robot movement and transformation, such as gears, chains, and so on. The 4D printing technology will provide a new technical solution for microrobots' movement and transformation. That is, the precise design and control of sensitive materials will be expected to replace traditional mechanical components. Gears are a good example. They could achieve the movement of robots, thereby considerably reducing the robots' weight and energy consumption requirements.

The fourth step is to reform the military logistics support process. 4D printing technology allows more weapon equipment to be manufactured in a folded form, reducing the volume of the equipment, facilitating long-range maneuvering, and decreasing the number of unnecessary damage that may occur during long-distance transportation. The 4D printed semi-finished components will have stronger malleable capacity and environmental adaptability, which can also reduce the variety and inventory of the equipment, improving logistic efficiency, and enhancing operational effectiveness.

4D printing has a profound impact on the military field. Particularly for some technological explorations led by the U.S., including the applications of body armor and ordnance, as well as facilities like camouflage techniques and outdoor communication systems – all will be affected by 4D technology.

Section 5 | Walking in Outer Space

4D printing has great importance in the military field. Also, it will bring major technological changes and breakthroughs to the aerospace industry. As humanity's footsteps in exploring outer space continue to move forward, the question of how we could travel among the galaxies more efficiently and in a controlled manner will become the direction of a new round of exploration in the scientific community.

Landing in space, exploring the future. A manned spacecraft is not only a ladder for humans to go into space but also a bond between space and humans – it carries countless possibilities and hopes. The traditional manned spacecraft we currently operate is often composed of two capsules. One is a sealed capsule for astronauts, which is equipped with a life support system that can guarantee their water and air supply, an attitude control system that controls the attitude of the spacecraft, a beacon system for measuring the spacecraft orbit, a parachute recovery system for landing and an ejection seat system for emergency life-saving. While the other one is the equipment capsule, which contains a braking rocket system that separates the manned capsule from the spacecraft orbit and returns to the ground, a battery that supplies electrical energy, an oxygen cylinder, nozzles, and other systems. The two capsules work together to complete the astronauts' round journey to space.

At the same time, because of the great difficulties in spaceflight operations, scientists usually take multiple measures on manned spacecraft to prevent and address some particular problems.

First, they use environmental control measures. Their main function is to adjust the temperature, humidity, and pressure inside the capsule and the spacesuit to ensure the amount of oxygen, ventilation, and water needed by the astronauts, and to absorb and dispose of waste.

Second are manual control measures. These are relevant mainly in the event of a failure of the automatic system, and when humans take over the operation and control the spacecraft to handle the emergency and avoid accidents.

Third, safe return measures. This is done mainly to ensure the safe return of astronauts. Other than providing a reliable heat protection layer to keep the return capsule from burning up, it is also necessary to make the braking overload device effective enough during the return process to guarantee that the astronauts' body can withstand the landing. Meanwhile, improving the accuracy of the landing point allows the astronauts to be located in time.

The spacecraft itself and the matched precautions are closely related to each other. More equipment and a more complex setup will undoubtedly be accompanied by an increase in the risk. 4D printing will be a new solution and hope for maximizing the simplicity and safety of the journey to space.

4D printing will enhance the astronauts' body functions. 4D printed high-energy food and drink could be implanted inside the human body, and along with the time and feedback from the human organism's needs, the food and drink will auto-release and dissolve in an orderly manner for the human body to absorb them. Besides, 4D printed microrobot doctor will be patrolling in the astronauts' body 24-hours a day to collect physiological information and to safeguard the astronauts' health and safety. Furthermore, there will be more 4D printed human-assisted functional devices that will mimic survival conditions in outer space and carry out environmental design and code programming accordingly. This will mean that the astronauts wearing these devices can move freely in the spacecraft and even outside in outer space.

When we are exploring space in the future, astronauts and scientists will no longer worry about the gravity unbalance as they do today. Under different atmospheres and gravity conditions, the 4D printed clothes will automatically adapt by changing their shape and size. Their shoes will also self-adjust for different landing contact surfaces to increase the grip or friction, ensuring that the astronauts could conveniently walk.

In the meantime, 4D printing technology will also make the manned spacecraft more able to self-transform. During the transport of astronauts to space, the function of the manned capsule will play the main role, while after the astronauts complete the space mission, the spacecraft will perform a function conversion, that is, the function of the equipment capsule will take over to send the astronauts back to the earth.

Not only that, the most significant difference between the 4D printed spacecraft and our Vostok-1 spacecraft is that the Vostok-1 and the launch vehicle are all one-time use spacecraft. They can only perform one mission. But, a 4D printed spacecraft and equipment will enable continuous reuse through their self-transformation and self-recovery mechanisms. On this basis, constant refinement and upgrading of equipment will be achieved.

4D printing technology and materials, besides being applied in spacecraft and aerospace, will also play an active role in the construction of space stations and spacewalks. With the upgrading of space equipment construction pushed forward by 4D printing technology, humankind will have the ability to step out of the earth and conveniently walk up the stairs to outer space.

With the increasing use of smart sensing technology in the materials for 4D printing, that is, the integration of smart wearable technology will bring tremendous changes and impact to aerospace and military activities. It will thereby build up our cognition and understanding of the unknown world. Tomorrow, with 4D printing technology's development and realization of its applications, humankind will walk towards the "universe village" from a "global village," opening a new world of unknowable future.

Section 6 | The U.S. Military's Efforts in 4D Printing

According to the news in the British *The Times,* 3D printers have allowed users to print out simple finished products at home, and scientists are now working on 4D printing. This means that they will use "smart materials" for printing, and the printed products could then automatically change shape.

Researchers are now trying to combine different kinds of plastics and fibers to create a "smart material" that can change shape on its own when exposed to external stimuli. Tibbits from MIT is conducting this groundbreaking research with a digital manufacturing company, which is the largest 3D printing technology company in the U.S.

As introduced by the researchers, 4D printing technology involves the application of special materials that change their shape when they sense motion status, temperature changes, or exposure to water, air, gravity fields, and magnetic fields. The fourth dimension mentioned here is referred to as the behavior of "self-assembly" of the materials. 4D printing is essentially 3D printing using composite materials, and in this way, a new function of transformation is added. This transformation can self-complete by being triggered at a specific time.

1. The U.S. Military Funding Support

As expressed by a professor from the University of Nottingham, the current 3D printing has become universal but slightly outdated, and it can be further perfected. Although 4D printing is still immature, researchers from MIT self-assembly lab believe that the technology will bring revolutionary changes to biological sciences, materials, software, robotics, manufacturing, transportation, infrastructure construction, art, and even space exploration.

For example, 4D printed car tires could enhance their grip in the rain, and flat metal plates or plastic could self-fold into the predetermined shape after being 4D printed. These properties could also be applied to automatic-assemble furniture.

Rosenberg, from the Nervous System design studio, wants to use 4D printing to create a dress that can be printed compactly and unfold automatically. By scanning the customer's body, the dress can be designed to fit perfectly and come with a series of inlay elements. The computer model of the design will then be compressed to take up as little space as possible to fit into a regular 3D printer. All the customers will need to do is to download the design, then, print it and unfold it.

Tibbits has been in contact with many companies since the first publication of his research last year. They hope to collaborate on 4D printing projects like sports shoes that can automatically adjust the support and ventilation based on different situations, water pipes that can automatically expand or contract according to the water flow, and even equipment that can mimic human's intestinal organs. In other words, Tibbits is currently exploring self-assembly applications that have larger forms and more complex dimensions.

According to a report in the *Science and Technology Daily* on 4D printing, Harvard University, the University of Pittsburgh, and the University of Illinois have received a US $855,000 grant from the U.S. Army Research Office to push the development of 4D printing further. They how to eventually be able to print out camouflage equipment that can change color according to the environment, and weapons that can self-assemble according to their environment.

Anna C. Balaz, the researcher from Harvard University, said, "We're proposing the development of adaptive, biomimetic composites that reprogram their shape, properties or functionality on demand, based upon external stimuli." The research team of Harvard University will focus on 4D printing materials at the microscopic scale. The research teams of the other two universities have already shown that they can 4D print larger objects, such as self-assembling pipes and other shapes, among others.

Even though 3D printing devices can print a wide variety of objects such as guns, toys, and other objects from materials like plastic, their weakness is that the printed objects need to be assembled. Meanwhile, 4D printed objects could self-assemble. Compared to 3D printed objects, 4D printed objects have a longer lifespan and can better adapt to the environment.

Besides, scientists believe that other applications of 4D printing technology include manufacturing of bridges that can self-repair when cracks appear, "adaptive pipes" that can expand and contract on their own, and many others.

5

LIFE

In a lab of Harvard University, a lifeless "flower," supported by the scientists' 4D printing devices and technology, magically realized the perfect self-transformation and blossomed. 4D printing technology not only gives plants specific characteristics of life forms but also allows new possibilities in the human body's functions, including gene sequencing, organ replication, as well as the improvement and optimization of various life forms. Furthermore, with the advancement of development and application of artificial intelligence, 4D printing will become a new manufacturing model for artificial intelligence devices.

Section 1 | Creating "Life"

In the dimension of time, giving the ability of self-transformation to the printed objects is the primary purpose and direction of 4D printing's initial stages. Even though 3D printing is not popular yet, it does not affect the pace of scientists promoting the development of 4D printing. A few days ago, in a lab at Harvard University, a lifeless flower was able to transform, and bloom on its own was created by the scientists' 4D printer.

This research focuses on advancing the development and application of materials. In the process of 4D printing, people use a variety of materials, including the hard and soft materials in a fixed form. Different materials combined in different specific ways achieve bending or transformation of each part of the structure under certain conditions. So, can we produce a similar effect with just one single self-bending material?

To achieve this, Harvard scientists use cellulose fibrils and wood pulp acrylamide hydrogels to develop a special gel. When the gel is extruded from the nozzle of a 4D printer, the fibrils are aligned with the extrusion axis, creating the hydrogel composite with directional properties. In other words, the composite can extend more easily in one direction. With the application of some preset mathematical calculations and encoding, after the composite is 4D printed, it can be made to stretch, bend, twist or curl under the action of the stimuli.

Section 2 | No More Withered Flowers

Scientists researching 4D printing materials have found their inspiration in nature. This was the complex structure of the natural morphological transformation, such as flowers' blossoms.

Scientists, therefore, mimicked the structure of plant cells. The aligned cellulose fibrils inside the plant cells limit their range of movement, so they mixed the cellulose fibrils extracted from wood pulp with acrylamide hydrogels (a gel that expands when exposed to water). When the composite was ejected from the 3D printer's nozzle, the fibrils were aligned in rows inside the acrylamide hydrogel, which means the printed object could only expand vertically but not horizontally. At the same time, based on composite guarantees, scientists have developed a mathematical model that can obtain curvilinear shapes from the crisscrossed designs and use the alignment of the fibrils to ensure that the curvature of the composite becomes consistent with the desired design (Figure 5-1).

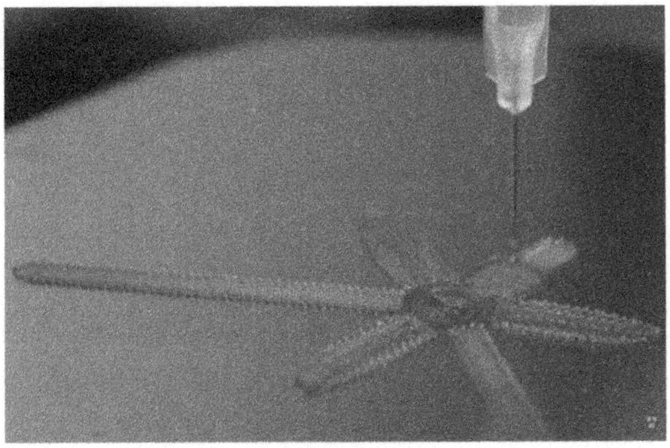

Figure 5-1 Printing a Flower

The result of the experiment is surprising. The two flowers were printed out in the same shape. When they were submerged in an aqueous medium, five petals curled in different directions. Besides, scientists also mimicked the design of a certain orchid. After the flattened-shape petals bent, the mimicked flower was very similar to a real orchid. After fluorescent dyes were added to the hydrogels of the composite, the flower was presented in a more beautiful form for people to appreciate.

In this way, the 4D printed flowers had the characteristics of life forms which made them very similar to the real flowers. With time, different growth patterns and characteristics will be further shown based on the action of different mediums, such as a flower as a bud, in full bloom, and overflowing with fragrance. Moreover, the mediums for the printed objects can be water, air, or sunlight, or anything else that could activate the 4D printing of flowers can be regarded as the medium. Flowers can even be freeze-frame frozen at their most beautiful moment or moved on to the next life cycle as needed by the people.

Section 3 | No More Aging Humans

With the successful experiments of making flowers bloom, scientists further worked on adding living cells to this basic composite to produce interesting living structures and even artificial organs. This process will be another breakthrough in the development of human tissue engineering.

At the primary stages, the medical science community will be able to 4D print drugs to control and treat the human body's diseases in the most timely and effective manner. For example, through the appropriate design of some polymers, a drug delivery vehicle can be created to deliver the drug into the human body.

The trigger medium to activate the drug in the human body is body temperature. That is, once the body's temperature rises to the critical point of fever, the drug contained within will be released and promptly control the disease. Concurrently, the drug vehicle without the drug will be excreted with the body's metabolism.

By the same principle, the medical science community can also design targeted inducing mediums according to the characteristics of its conditions, so that drugs could be delivered to the corresponding locations in the human body, and once the pathological symptoms appear, the drug could be triggered and released. Thus, diseases could be perceived before they show symptoms, and the most timely medical treatment and rescue plan could be achieved.

When the prevention and treatment of diseases are realized by 4D printed drugs, the future of human science will be further committed to 4D printing in R&D and the application of technology for human organs. It will thereby enter a new era of "organ replacement" and opening up new life channels. According to the latest news, a team of experts from Xijing Hospital of the

Fourth Military Medical University, in conjunction with a resident national key lab, adopted 4D printing technology to successfully treat and save a 5-month-old infant who had congenital heart disease with bilateral trachea severe stenosis that was life-threatening due to potential further tracheal damage. The 4D printed external tracheal stent tailored to the shape of the infant's trachea was made of degradable material, which not only met the performance requirements of shape, stiffness, strength, and elasticity of an external tracheal stent but also conformed to special requirements for human biocompatibility and biodegradability. The stent could be gradually degraded and absorbed by the human body in the next two years, saving the infant the pain of second surgery to remove it. The infant's tracheal tube was removed on the same day after the surgery, and on the fifth day, cardiac ultrasound and chest CT scans showed that cardiovascular malformation was completely corrected and the bilateral trachea recovered well. This is the first time in the world that a 4D printed external tracheal stent was successfully used in the treatment of an infant with complicated congenital heart disease with severe bilateral tracheal stenosis.

In the future development of science and technology, through 4D printing, what humans can replace will no longer be limited to the diseased organs, but also the weakened, aged organs and also auxiliary devices for human functions. Perhaps, there could be a new drug to keep humans young forever and thus compose the legend of humankind's immortality. Besides, the scientific community has already made initial attempts in the lab to grow cells in a flat framework and then for them to subsequently change the shape of the tissues.

Section 4 | Implantation of Human Organs

In recent years, with the development of technology, 3D printing has taken the lead in gaining application breakthroughs in the medical field. This is mainly due to the significant demand for personalized customization in the medical industry (especially in the field of restorative medicine) and there is a rare standard of quantitative production. At the same time, small batches, high precision, as well as customizable printing based on individual differences are the advantages of 3D printing technology and are also applied in various medical fields, including the printing of limbs and organs. However, compared to 4D printing technology, it is still relatively primitive and outdated.

The best description was published in an earlier report in the *International Herald Leader.* It has the following views on the applications and changes brought by 4D printing technology in the medical field. According to the report, 4D printing technology not only plays a role under extreme conditions but it can also make a big splash in a tiny space. For example, inside the human body, 4D printed objects become new types of human implant materials that could even replace organs.

Using its automatic shape-changing features, 4D printing technology has unlimited potential in the biomedical field. Some biomaterials' companies are already involved in the development of biological heart stents with memory functions. The conventional heart stent in the medical field is usually made of memory metals, which will automatically hold open and dilate the blood vessel passages after implantation in a predetermined position through a blood vessel. But the problem is that the metal stent is not degradable, and unless it is manually removed, it will remain in the body forever, causing many complications and adverse effects.

If a new biological heart stent is developed using 4D technology, it will be able to transform as well. Firstly, it could enter the human body in an extremely tiny form, and when the condition in the body would be most suitable, it then would expand and transform into a hollow stent that could hold open. Since it will not be made of metal, it may also be degradable. After the drug-coated stent completes the mission of dilating and unblocking the blood vessel at its predetermined position, it will automatically degrade in the blood.

Similarly, the use of 4D printing technology to print out the replacement of human organs, such as joints and cartilage, will have more bionic qualities and will be able to make itself more similar to the real organs of the human body.

Section 5 | Bioprinter

Anthony Atala, a pioneer in U.S. surgery and regenerative medicine, once gave a speech, which was about the rising health crisis because of the increasing prevalence of human lifespan extension and organ failure. In the U.S. alone, as many as 100,000 people are waiting for an organ transplant, and that number doubles every decade. Yet, against this backdrop, the number of organs available for transplantation has not increased and is less than one-third of the total number of people who need organ transplants. On this occasion, on the stage, Atala demonstrated a special 3D printer that could create a prototype of a human kidney. When Atala was wearing medical gloves holding a complete kidney that has just been "printed out," the audience was deeply shocked!

If the kidney printed by Atala was still in the demonstration stages, the cardiovascular surgeon at Miami Children's Hospital using 3D printing technology was able to make a girl's heart with real-world application. A 4-year-old girl named Adanelie Gonzalez suffered from a congenital disease called total anomalous pulmonary venous connection (TAPVC), which meant her the veins' were unable to deliver blood to the correct part of the heart, resulting in breathing difficulties, lethargy, and frequent illness (weak immune system). After several restorative surgeries, doctors realized that without a permanent repair, Gonzalez's life would be only a few weeks long. At that point, doctors successfully performed the surgery using a 3D printed donor heart. Now, Gonzalez's blood is flowing normally and she is currently recovering at the hospital.

Meanwhile, 3D printing has gradually replaced infected jaws, cancerous vertebrae, and deformed hip bones in the world's leading medical community through the human anatomy. In June 2014, Len Chandler, who is now a 71-year-old from Rutherglen, Australia, was diagnosed with cartilage cancer in his heel. His surgeon contacted the experts from Australia's Commonwealth of Scientific and the Commonwealth Scientific and Industrial Research Organization (CSIRO) and they brought expertise from an Australian company, Anatomics, which produces and customizes medical equipment. By using the heel-bone anatomy chart provided by the company, the research team single-handedly developed an ideal implant and provided it to the surgeon from St Vincent's Hospital. The important thing was that surgical sutures could be inserted in the holes of the implant, which connected the smooth and rough surfaces of the bone to allow the tissues to grow closer together. In July, a titanium heel bone implant (Figure 5-2) was successfully

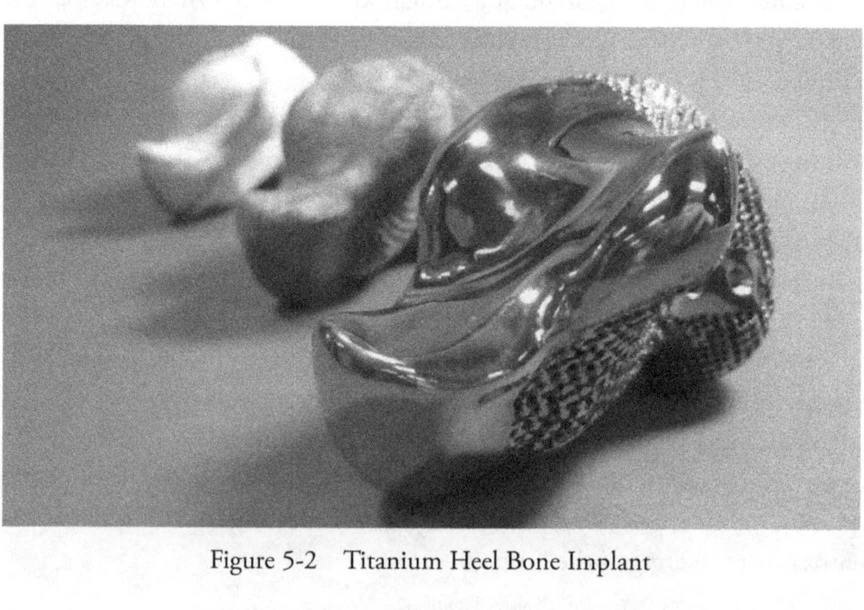

Figure 5-2 Titanium Heel Bone Implant

implanted in Chandler's foot. Three months after the surgery, Chandler returned to the hospital for a follow-up visit and the doctor said that he was recovering well and his heel was able to bear some weight.

Internationally, "bioprinters" have been used in many applications. In China, similar success stories are common. Doctors at the Peking University Third Hospital successfully implanted a customized 3D printed vertebra to a 12-year-old boy to replace the segment of the boy's cancerous vertebra in his neck.

During the 5-hour surgery, surgeons removed the cancerous vertebrae and implanted a 3D printed vertebrae between the boy's first and third vertebrae. The surgery involved removing nerves, carotid arteries, and cancerous tissue from the spinal cord and fixing the artificial vertebrae with titanium screws. According to the director of the hospital's orthopedics department, "If using existing techniques, the patient's head needs to be framed and pinned after the surgery, which means the patient's cannot touch the bed while resting, and this lasts at least another three months. But with 3D printing, we can mimic the shape of the patient's vertebrae, which is a lot more powerful and convenient than traditional methods." The biggest advantage of 3D printing is the ability to customize medical implants to create artificial bones that perfectly match a patient's skeleton. Currently, the boy is in good health and his recovery is progressing as expected.

The "bioprinter" can cultivate the patient's cells and then use 3D printing technology for replaceable skin or organs, realizing its great potential in some areas of medicine. Moreover, the development of 3D printing will eliminate the need for organ donation waiting lists. Likewise, skin transplantation will become history. Along with the application of 3D printed human tissues in drug testing, it will also reduce the need for animal experiments.

With the development and application of 4D printing technology, bioprinting will be further perfected. People could cultivate the printed cell layer directly on the wound, and it will transform into a shape as the wound evolves. Or, an instrument could even be penetrated deep into the patient's body to detect the damaged cells and replace them with the new printed cells through a minimally invasive operation. Instruments could use their probes to repair wounds directly.

Taking bioprinting from imagination into reality allows the human body to timely renew cells and promptly repair and renew the diseased and aging human tissues. When people frequently use "face printers" to pursue eternal youth, a guy who sees a sexy girl may say, "Hey, girl, you look good!" and the girl may turn around and tell him, "Thank you, young man, I've been printed for 87 years." The young man would then look back with a light smile, "I've been printed for over a hundred years."

Section 6 | New Anti-Cancer Technology

According to the *World Cancer Report* released by the World Health Organization in 2014, China topped the list with almost half of the world's cancer incidence rates.

The report predicted that by 2030, the number of global cancer patients will increase by 50%. The report believes that the global cancer incidence situation is grim, and the morbidity and mortality of cancer continues to rise, especially in developing countries. In 2012, half of the new global cancer cases occurred in Asia, and most of them occurred in the mainland of China. Among the global cancer cases, lung cancer has the fastest rise, which is closely related to the increase in tobacco sales. Globally, lung cancer is the biggest killer, accounting for 19.4% of cancer deaths. It is followed by liver cancer, accounting for 9.1%, and gastric cancer, accounting for 8.8%.

Presently, we have almost no effective medical technology for cancer. For example, the most common chemotherapy technology, although it does have certain effects in eliminating the cancer cells, it also eliminates the immune cells at the same time. But, 4D printing technology may change this situation.

Through some of the previous chapters in this book, we received some understanding of 4D printing technology. The entry of 4D printing technology into the medical field, or the field of printing cells, will bring a fundamental revolution to the current medical technology. That is to say, we can mimic white blood cells and 4D print them with bio-materials. Of course, the key to this lies in printing materials, which will have great uses in the medical field.

We can simply understand that a 4D printed cell "robot" is an anti-viral nanorobot made from the materials that are designed according to DNA strands, in the form of a clamshell or the same as a normal human cell. When such robot cells are input into the blood and come into contact with specific cancerous cells, they will release specifically targeted antibodies that halt the cells' growth or mimic the behavior of a specific form that attacks the specifically targeted cells.

The medium that triggers the cell robots is the cancerous cells. The triggering medium could be any specific kind of cancerous cells or a variety of cancerous cells. While the 4D printed robot cells are flowing in the blood and patrolling the human body's white blood cells, once they detect the cancerous cells, the robot cells will be able to target and attack and exceeds the killing power of the white blood cells (Figure 5-3).

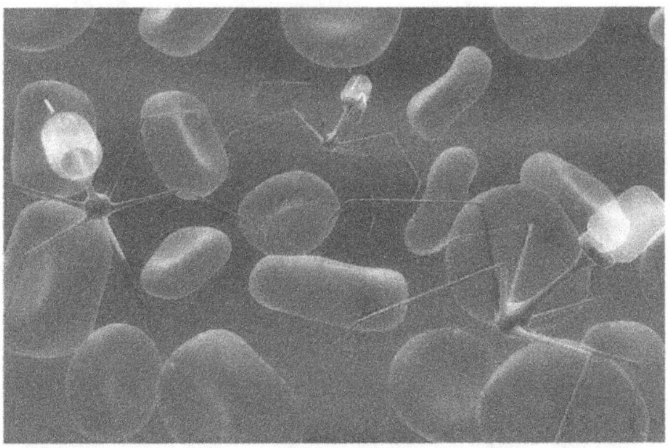

Figure 5-3 Attacking Cells

For patients who have had cancer, we can inject the 4D printed targeted robot cells into the cancer site to attack or devour the cancerous cells. This will effectively improve the current treatment methods for cancer and provide multiple new solutions for humans to overcome cancer.

Solution 1: 4D printed "guards" for the human body

With the support of nanotechnology, 4D printed non-therapeutic nanorobots will be able to function as the guardians of the human body and conduct 24-hour patrolling inside the human body.

On the one hand, the "guards" can timely clean up the residual "garbage" in the body, especially in the blood vessels, and the "garbage" could then be excreted with metabolism. On the other hand, which is more important is that the "guards" can also detect cells with cancerous potential in time, and issue a warning immediately or directly kill the cells before they turn into cancerous cells to ensure the stability and harmony of the human body.

Solution 2: 4D printed anti-cancer drugs

4D printed objects with the ability to self-transform have unlimited potential in tiny spaces. In the process of cancer treatment, we can replace or repair damaged cells and tissues with 4D printed organs or stents.

Taking the biological heart stent as an example, the new 4D printed biological heart stent could enter the body in an extremely tiny form, and then expand and transform into a hollow stent that could be held open under a suitable environment in the body. The stent is like the 4D printed underground water pipes that can adjust the capacity and flow by expanding or contracting the radius of the pipes. It can even self-repair when damaged or decompose when "scrapped."

Solution 3: 4D printed human skin

Scientists in the Netherlands have recently actualized the use of stem cells as ink for 3D biological printing of human skin, which is regarded as the prelude to 4D printing of human skin. Relying on the self-transformation characteristics of the editable materials, 4D printed skin that fully integrates the features of individual skin textures will achieve the greatest degree of fit in the process of replacing the cancerous or burnt human skin. This can dramatically improve the healing rate of carcinogenesis or other dermatological patients.

Solution 4: 4D printed living tissue implants

Earlier, researchers in Melbourne have found a way to generate a kind of cartilage that can "self-generate" for cancer treatment and replace damaged cartilage. On this basis, scientists can use editable materials and incorporate them with the genetic parameters of the human DNA strands while editing. Upon that, the living tissue implant obtained by 4D printing the materials will minimize the rejection by the human body. In case of adverse reactions, the implant can also self-transform in line with the actual environmental conditions in the human body to come out with the best solution to adapt to it.

Solution 5: 4D printed medical equipment

The process of cancer treatment is a life-saving process but also a life-threatening one, at the same time. This is because modern medical treatment kills many healthy cells that are useful to the human body in the process of killing cancerous cells through radiotherapy. In this treatment process, if cancerous cells can be isolated and the radiation area can be more precisely located, the cancerous cells can be killed effectively without damaging the human body. This will help in improving the success rate of cancer treatment.

Medical equipment like 4D printed radiotherapy auxiliary will play an active role in this aspect. These 4D printed medical equipment machines can enter the human body in small volume and transform according to the living environment of different parts of the human body to effectively isolate cancerous cells and protect the healthy areas so that the cancer treatment becomes "harmless." This is especially important for tumor treatment of some vital organs or vulnerable areas, such as the nose, eyes, ear, etc.

Implanting 4D printed objects into the field of biomedicine, particularly expanding these to human body applications, is the gospel for the development of human health and medical treatment. However, even though we are excited by the prospect, we must also be calmly aware of the risks involved and we must not underestimate them. Let's say the 4D printed cells or nanorobots wandering around in the human body can easily turn into prototypes of biological weapons used by people with bad intent if the monitoring is not strict. The self-transformation feature of the editable materials makes it even riskier for use by criminals compared to 3D printing. For example, forbidden objects like guns. The 3D printed guns are concrete and direct, which is relatively easy to detect and control, while guns that are printed via 4D printing can take any form, and only under certain environment and action of the triggering medium, they will transform into their predetermined shape. This is unpreventable.

Section 7 | Human Genes

Genes, as one of the most mysterious mysteries of humankind, have always been one of the most critical factors in human development because they contain the unique password to each individual. Therefore, solving the mystery of genes has become humankind's never-ending scientific quest, in hopes to realize the optimization of genes to better eliminate diseases, prolong life, improve intelligence, and beautify our appearance, among other factors.

Through the continuous exploration of 4D printing, humankind has found a broad road of solutions for decoding genes. In the future, we no longer need to complain that our inborn conditions are lousy, and this will not be an excuse for humankind anymore because the genes of each individual would have been "modified" from the very beginning. They will be completely 4D printed following humankind's optimal standards.

At present, a company named Cambrian Genomics, a division of Singularity Incubator, has created the world's first DNA laser printer capable of producing the world's first synthetic gene. It has already allowed scientists to ensure that children do not acquire the genes that will cause future conditions such as breast cancer, cystic fibrosis, and Huntington's disease through genetic modification and repair. On top of that, 4D printing technology can further kill the uncertainties before they turn into reality. This is done through a program preset at early stages, when there is a loss of genes or when genes are mutated. The "ambush" medium will be triggered to "maintain" or "reset" those genes. In this way, the countless people who are currently suffering from various cancers will be completely freed from such nightmares.

Section 8 | Artificial Intelligence

Nowadays, artificial intelligence has become the focus of our attention in the science and technology industry. In the popular TV series *Westworld,* we have seen a new version of the human living space, and the AI hosts serving us there seem to be different from ourselves, even from the way they were born.

The first scene of the series shows the printing of the AI hosts. In this "Westworld," which provides people with travel and entertainment, all its hosts are born in this way. This scene indicates a new movement of life-making is on its way.

From this moment on, the method of new humankind's birth will no longer be through the pregnancy of the mother, but the new printing technology that is based on the scientific and technological development, which currently consists mainly of 3D printing and 4D printing. Besides, the generation and development of the new humankind's consciousness and thinking will no longer rely on the traditional teaching and learning, but the writing, inputting, and upgrading of program codes. From this process, we not only appreciate the "mind-blowing" abilities of 4D printing artificial intelligence but also witness the comprehensive embodiment of future technologies. These include the development of new technology programs, the applications of Internet cloud technology, among others.

The 4D printed artificial intelligence is arguably the most comprehensive application of humankind's future technological development. In the previous chapter, "Product," we have received some knowledge and understanding of 4D-printed artificial meat. To make the meat, cell bodies obtained from small living tissues became one of the raw materials for printing after they were supplied with nutrients and cultivated in vitro.

For 4D printing artificial intelligence, things used for its creation, other than the metal devices, are more of the new organic materials. The use of these components will give 4D printed artificial skin, tissue cells, musculoskeletal, and other parts, more "life-like" characteristics.

For example, the 4D printed human bones and cell tissues which are originally lifeless machine-printed objects would now have the ability to grow over time. At the same time, the "embodiment of life" implanted in the body of artificial intelligence triggered by external media will produce life signs like "injury," "bleeding," and even "death." Artificial intelligence will thus become a new kind of "human replication." In particular, with the breakthrough in the technology research of liquid metal, the flexible body characteristic of artificial intelligence will be further encouraged.

6

COMMERCE

The rapid development of science and technology will inevitably bring about changes in the existing modes and business forms of human society. So, what kind of changes will the advent of 4D printing and the advancement of its materials and technology bring to our social ecosystem? What will be the variables for the structure of today's widely used economy models like B2B, O2O, P2P? Especially for those "ready-to-go" new sharing economy models, will they be able to join the 4D printing movement and be further boosted? Variables are everywhere, and the only thing that will not change is that 4D printing is about to open up a new model of the human business ecosystem.

Section 1 | The Business Ecosystem Revolution

I have been thinking that with 4D printing as a new production means, our life will experience such strong disruptions, that I wonder, what kind of a business scenario will the future bring? The current hot Internet Plus and the shared economy model represented by UBER have brought some profound changes to our existing business scenarios, especially to the huge business entities built in the last commercial wave. Before, we were in a period that was relative to the material shortage and technology was not as advanced as today's. Lenovo, Haier, Suning and other huge business entities that rely on their powerful offline channel systems built a commercial aircraft carrier in their respective fields before the era of Internet Plus.

But when it comes to the era of Internet Plus, some of the key elements that once led to their success have now fundamentally changed. In a fundamental sense, the Internet does not appear as a new production means but rather exists as an information tool. However, the in-depth social changes it has brought about have transformed our current business ecosystem entirely, and more changes are happening on the edges of this ecosystem. Therefore, in the era of 4D printing, what has been changed is not information tools, but the entire production means. Just as the emergence of the steam engine has changed the entire production means, production tools, and production methods, 4D printing also appears in our time in the form of such a new production means and production methods.

We can imagine that once the production means, production tools, and production methods are changed, the transforms they bring will far exceed the impact of the Internet Plus wave we are facing today. A simple example is that before the steam engine revolution, what were the production tools we relied on to sustain ourselves? It was nothing more than the initial hand tools of the traditional agricultural society. So, the business ecosystem of the entire society was built on a relatively slow production model at that time. After the industrial revolution, due to the changes in production means, production tools, and production methods, the balance between supply and demand for goods was broken, and thereby we saw the emergence of a new business ecosystem, which lasted for a long time before the Internet.

Upon entering the era of the Internet, although it did not change the entire production means as the industrial revolution did, it has changed the information flow of production means, production tools, and production methods. Judging from the current situation, the information flow has transformed the structure of the entire business ecosystem. So I'm thinking, with the emergence of 4D printing technology, the series of changes of production means, production tools, and production methods that it brings will be similar to another new industrial revolution as the one triggered by the steam engine. Under the new "industrial revolution," our business ecosystem and business structure will unavoidably undergo fundamental changes.

It is foreseeable that in the era of 4D printing, business forms such as Alibaba, Wanda, and the current hot O2O model will be replaced, and then reappear and exist with a new or unknown business terms. The huge factories and the powerful assembly capabilities that we are proud of today will be replaced and exist in a very embarrassing state during the era of 4D printing.

If 3D printing changes the manufacturing ecosystem, then 4D printing will change the entire business ecosystem. A simple example is a change in IKEA. In the future, when we have a 4D printer and we can print out several layers of materials at home or in a specific place, we will just need to put them on the balcony and shower them, and after a while, a desk will automatically assemble and present itself in front of us. We will no longer need to assemble it by ourselves. Therefore, 4D printing will bring not only a revolution in materials but also a revolution in the business ecosystem.

Section 2 | New Model of Business Scenarios

In the era of 4D printing, most products are not going to be produced in factories, instead, they will be produced by different 4D printers equipped in a community or several communities. Regardless of the current shopping malls or e-commerce platforms, the value and significance of the current business methods will be lost. For general products, desktop printers for home-use will be installed in thousands of households. Users could then print the products they will need based on their ideas. Moreover, they could also print out some of their ideas into products and use information tools to complete transmission, transactions, etc. The role played by an individual user in the 4D printing era will be diverse. They could be consumers, users, and producers.

For some special products, especially those that are relatively large-scale, the printers used will be provided by organizations or individuals, and consumers will only need to put forward their needs or ideas on the big data platform of the Internet of Things (IoT),[13] the platform will then analyze the consumers' preferences and then provide them with the optimal solution. Hence, in the future 4D printing era we are waiting for, what kind of words can be used to describe the business ecosystem? At least the currently existing business terms are not accurate enough. Perhaps we can only get a sense of the new business terms when we enter the new era.

Just imagine, today we keep spending money on platforms to create the so-called "new monopoly business systems." Why don't we invest in this disruptive technology that will truly disorder the future business ecosystem? This is the future. Based on powerful technologies on the way, such as smart wearables, artificial intelligence, and the entire IoT era established by them, the flow of our information will be further transformed. In addition to the changes in the entire production methodologies brought about by 4D printing technology, which I cannot simply describe in words, all I can say is that today's entire business ecosystem will be overturned much the same way it was overturned in the past.

Section 3 | Community + Crowdfunding + Sharing

The future may be nearing us through ways we currently can't recognize, but if we have to find a few terms from the existing business terms to help us understand it, I will choose community, crowdfunding, and sharing.

In the days of 4D printing, like-minded friends or neighbors gathering together will be called a group according to the popular term of the day, which doesn't have to be as big as today's Internet-based community. It may be just a few people who have a common demand, for example, they are all interested in innovative home products, and these people constitute the group. Then, they could put their funds together, which is called crowdfunding, to buy a 4D printer that is suitable for printing innovative home products.

After that, it's all about sharing, that is, the crowdfunding group members can use this 4D printer according to their needs, and they only need to design the product they want to print on their intelligent terminal device. This will not be limited to a PC but it could extend to the smart wearables, and then the design could be transmitted to the 4D printer for printing. This operation will be the same as the text printer today. Besides, the group can also rent out the idle time of this printer on an Internet platform, thereby forming a bigger shared economy circle.

Section 4 | Consumer Safety Index

A car enthusiast is longing to produce a user-customized steering wheel by 3D printing it in a basement workshop. Although the simply-printed steering wheel does not undergo quality and safety monitoring, its design file can be easily bought on a popular website that sells novel car accessories, such as novel air fresheners and gearshift handles, and others.

Imagine a car hobbyist sold a printed steering wheel online and its buyer then installed it in their car. A few weeks later, while driving at a high speed, the buyer found that when they made a sharp left turn, the user-customized steering wheel disengaged. However, it was too late and the car crashed, killing the driver.

Yet, the steering wheel made by 4D printing technology can avoid this tragedy. Just as the water pipelines that will naturally transform in response to changes in the external environment, the steering wheel will spontaneously form the most suitable and seamless connection with the car according to the car's conditions. Ultimately, it will interact and integrate with the car, eliminating the possibility of disengagement. Moreover, if the car itself is 4D printed, it will also make the corresponding transformations in the event of a car accident to ensure the safety of its passengers.

Section 5 | New Logistics Forms

Driven by the concept of globalization, the global village has made the circulation of goods among countries and cities more common and frequent. In particular, e-commerce companies represented by Jack Ma and Taobao have injected a booster for the transportation and circulation of goods. However, the increasing scarcity of energy and resources worldwide, such as petroleum and coal, will greatly slow down the pace of physical goods transportation and circulation within globalization. After all, without the support of energy and resources, globalization and the global village may fall back and become a beautiful vision once again. This means that the scope of physical goods' transportation and circulation will gradually shrink and localized production and manufacturing will likely to also once again become the mainstream.

At this point, the traditional concept of globalization will evolve into a combination of digital modeling globalization and manufacturing localization. A "supplier," or a designer, will provide the design program with a product's printing solution on a computer and will store it as a digital model. Then, they could sell the design program to the "demander" in any part of the world, and the transaction could be done via the Internet. After receiving the design program, the "demander" can print out the physical product via the 4D printer in their location, thus realizing the localized "production and manufacturing" of the product.

Such a convenient way of transaction and product manufacture sounds like it is from sci-fi. In fact, a sharing website called Thingiverse, where designers can upload their digital works for others to download or print out has become a reality, and there are already many people who regularly share files, pictures, and videos online.

The magic of 4D printing goes far beyond this. Relying on the changes in the parameter settings of the time dimension in 4D printing, the locally printed products will also be able to make adaptive changes according to the physical environment of their locations. This could be the so-called "tailor-made" aspect of the design and it could be difficult to achieve when products are produced in a unified production base within the traditional model of globalization.

In the era of 4D printing, printing objects will realize the storage and transmission of digital information in the physical world through computers and the Internet. The problem of long-distance physical goods transportation constrained by the lack of energy and resources will also be resolved. As such, waste of energy and resources will be minimized to improve the efficiency of goods transmission.

7

THE MAKER

There is an old saying, "There is but one secret to success: never give up." The saying is also incisive in the development of personal 4D printers because people who are pushing 4D printing out of 3D printing models are titled, the "Makers."

Section 1 | New Era of the Makers

The Makers refer to a group of people who like or enjoy creativity. They pursue to realize their creativity through DIY projects, which means they like to make things by themselves and turn various ideas into real products. They are "super users" of software companies, and also a bit like "hackers," they modify the software and improve technology according to their own will and expectations. Besides, when the Makers advocate personalized customization, they also advocate open-source sharing. This reflects self-confidence, whereby they believe that even when all technical details are free and open, no one in the world could do better than them in this small field.

Even though the "Makers" are just sprouting in China, they have a long history of development abroad. Steve Jobs is one of the pioneers in the world of the Makers. He's been obsessed with innovation since he was a child, and he once tried to change the pulse frequency of the telephone to make free calls. His persistent pursuit of innovation makes Applewhat it is today. Furthermore, the diversity and sharing in the Makers' culture make it inclusive and full of affinity, allowing more people to join in. Star Simpson, majoring in Electrical Engineering and Computer Science at MIT, was addicted to open-source hardware. In her opinion, there is no difference between low tech and high tech, and they both are tools used to bring our ideas to life.

With the development of the Makers' culture, a lot of novel technologies and products were born, and the traits of the Makers come to the fore. The first is that they must do the work themselves. The second is that they must share the novel stuff they've made, which means sharing thorough design drawings and source codes with everyone. The third is that they have to take advantage of network communication to work with like-minded people to discuss and improve products' creativity positioning, technical difficulties, financing methods, profit models, marketing methods, among other factors. It is their enthusiasm and selflessness that breaks the technological blockade of large factories, large companies, and multinational corporations, and lays a solid foundation for the arrival of a new era of "personal intelligent production," "family intelligent production" and "network community intelligent production."

Section 2 | 4D Printing for Creativity

While people are still exploring B2C and discussing C2B, the applications of 4D printing technology based on 3D printers have been weaving another technological blueprint for the future world – a unique and personalized world showing independent creativity. In this world, all people create and everyone is an artist. Everyone's items are unique and exclusive to their possessions only. Every household can set up its factories to break large traditional manufacturing enterprises into pieces. What's more, the 4D printed personalized products will transform into infinite possibilities with the changes of the time dimension and living space. Let's imagine that the items we use can transform with the changes in the existing space and triggering medium. The clothes we wear can transform with the growth of our body and the changes in our composure. Even the organs in our body can constantly repair, improve and upgrade as the living conditions change. All these ideas are possible with the advent and development of 4D printing, which will make our future world magical and creative.

3D printing has led product customization to cross the high-cost threshold, and 4D printing has further lowered down the cost for the personal customization of products. It has enabled it to move forward without cost constraints and reach a rapid upgrade.

The basic requirement for the popularization of low-cost product customization is there the sharing of the design program of digital products.

Once, we praised design software that shaped our world because there was a computer design file behind almost every architectural model, every product prototype and every finished product. This includes the chair you are sitting on now, the book you are holding in your hands, the computer on your desk, the car you drive, the house you live in and even the buttons on your shirt. All these things were digitalized before they were produced. In short, the design file is the language of modern engineering.

Before, and even after, all the products are produced, design files are often not made public and kept secret. Because they are not only the fruits of designers' labor but also the expression of their wisdom. They are always "protected" under "intellectual property."

However, with the promotion of 3D printers and the development of 4D printing technology, the Makers have come up with more design programs through their creativity. Even more and more individual Makers and Maker groups are emerging, thereby more and more creativity and products are coming to life. Because of Makers' self-confidence and their advocacy of resource sharing, every "eager" heart in the world can enjoy the "welfare" of design documents. As long as people have a good intention, they can easily obtain the original programming codes of product design released by the Makers from the relevant media or network information platforms. Thus, they will take part in technical communication and learning interaction at the same time.

Furthermore, everyone can partially modify, adjust, and redesign the products according to their preference based on the original design by Makers. After fine-tuning some of the code and programs, new product design codes can be fully incorporated with your aesthetics and preferences. Then, the product becomes your unique and exclusive "personally customized" product in the real world. By promoting such practice, as long as someone wishes to do so, they can create a personal customized product on the fertile resources established by the Makers.

A 4D printing movement of "everyone's creative ideas and everyone designs" has been brewing and fermenting, and it slowly set off the prelude to a new production model of the manufacturing industry.

Section 3 | 4D Printing for Civilians

As the saying goes, "You can know the distance from the near, the more from the less, and the salient from the subtle." We can see that from the efforts of the Makers, soon, right after 3D printing, 4D printing will enter all aspects of our future life and work.

At the second White House Science Fair on February 7, 2012, a group of Makers was invited to the White House State Dining Room to demonstrate the innovations from the public. Joey Hudy, a 14-year-old Maker, with the help of the U.S. President Barak Obama, fired the "extreme marshmallow cannon" that he invented. In 2012, in Amsterdam, the capital of the Netherlands, a modern art exhibition focusing on the latest achievements of 3D printing technology kicked off, which made us fully feel the infinite power of the public's creativity. As a great man once said, "The people, only the people are the driving force to create world history."

With the combination of 3D printing and 4D intelligent data, the threshold of product design and manufacturing almost becomes zero. By facilitating ordinary users to easily group themselves as Makers, it will be possible to kickstart the miracle of the new industrial revolution and move the expensive 3D printers from high-end factories into thousands of households. Additionally, the development of digitalization will usher in the first full-blown period of "personal hero opportunity" in the history of human society because we are all born Makers since the very first clay figurine we made in our life as a child.

If things go on like this, people will continue to upgrade their fantastic ideas and apply the tools which are already advancing in our daily life, and thus promoting the overturning of the existing technologies and starting the revolutionary waves set off by the new technologies. The promotion of 3D printers as well as 4D tech, which is an upgrade on the technical level, will help to move humanity past the castle of the "high-tech" and into the homes of ordinary people. All of this will be pushed by the interest and efforts of the Makers.

Section 4 | Entrepreneurship Started by a Printer

The popularity of 4D printing has turned everyone into an inventor full of creativity.

With the promotion and applications of 3D printers and 4D printing technology, we can not only print the needed personalized items at home but also easily start a business as a manufacturer of 4D printed products. As long as there are needs, whether models, parts, or even food or cars, we can print creative and personalized products for them through a click of a mouse.

Similar to the agricultural era, the original one acre of land was self-sufficient to support a family. But now, a limited cost of a printer allows everyone to easily set up their own "manufacturing" factory and embark on a "production" entrepreneurial road. Such an enterprise will minimize the costs of labor, materials, and finance for 4D printing manufacturers. This means that an individual plus a printer (including materials) will be all that's needed. For families living in a neighborhood, they could divide the work to meet their different needs. All these can be called manufacturing by all. Different from the ancient arable and weaving, today we see a prosperous scene of a Chinese society where every household is producing and manufacturing goods to fulfill the material needs of the society, science and technology. As we are relying on the advancement of science and technology, we are not using hammers and hoes, but the high-tech 4D printing equipment and memorable polymer materials. What we are producing is not just grains, but various items that meet the needs and aspects of our life and work.

Clothing. In the 4D-printing manufacturing process, households in charge of clothing production and manufacturing will create a real "personal customization" history of human clothing as well as enter a new era of "one piece per person." As such, only one piece of clothing will have the ability to self-transform into various forms. Firstly, it will transform into different shapes and colors according to the wearer's needs on different occasions. Secondly, based on the contact with the human body, it will also adjust its thickness and materials with the change of the human body's temperature to meet the wearer's needs of wearing the same piece during all seasons. In addition, it will even be pre-programmed with the needs of the human body's skeletal growth, growing together with the wearer. Every piece of such clothing will truly realize the "personal customization" for the clothing manufacturing industry.

Food. The 4D printing of food will change the bad habits of people's traditional diet and we will be able to say goodbye to "junk food." Firstly, people will be able to keep the good and discard the rest in the production of raw materials, which largely break the current predicament of "illness entering through the mouth." Secondly, the enhanced matching between food ingredients and human body functions will avoid the negative impact on the human body and the food waste caused by "reckless eating." Thereby, it will truly realize the "personal customization" of diet.

Housing. Regarding the 4D printed houses, we have already introduced the houses fixed in a certain place in other chapters. Here, we are going to dream freely about mobile residences, which will play an important role in the traveling life of people pursuing the future. Firstly, a 4D printed mobile residence will be set with the user's personal preferences. Hence, its transformed form will meet the user's individual preferences to the greatest extent. Under the action of the medium, it will transform into a new-style tent for users preferring tent form and transform into a recreational vehicle for users preferring recreational vehicle form. Secondly, the 4D printing raw materials used in the manufacture of mobile residences will have greater shrinkage and curling capabilities. They will also be able to recover under the action of the medium to achieve multi-repetitive use.

Transport. As a representative of transport, we have introduced 4D printed BMW in a separate chapter. In the future society, more convenient transportation means perhaps just a pair of shoes on our feet, just like the Wind Fire Wheels on Nezha's feet.[14] Maybe a chair under our butt could be a new driverless convertible car, or even a hat or an umbrella over our head, could take us flying with just a single click. All these incredible types of equipment will come along with the continuous advancement and development of 4D printing technology, the development of raw materials performance, and the program coding presets. Plus, the popularization and applications of science and technology development to all these factors will also realize these dreams.

Section 5 | The Source of Power for 4D Printing

In the 21st century, each of us wants to use unique products, just as each of us is unique. This makes us believe that the market is full of endless business opportunities, and personally customized products are lucrative. Makers are forming a new industrial organization model, which is an interest and project-oriented driven company on a smaller scale with tendencies to be virtual and informal. The Makers team up and regroup their operations. The team members in the company are far fewer than the traditional large companies, but their innovation ability is much stronger. Makers not only perform well in niche markets but also frequently set off huge waves in mass markets. Thus, the existence and development of the Makers will become the lubricant and accelerator for the development of 4D printing.

If we go back in time, if it weren't for Steve Jobs, it's hard to imagine that Apple would have grown into the world's most valuable company. When Jobs left, Apple weakened near to bankruptcy. When Jobs came back, Apple prospered. First of all, it took an iPod music player to make Sony's Walkman extinct, which had dominated the world for many years. Then, it used the iPhone to make Nokia, the world's No. 1 mobile phone company, announced that it would never make mobile phones again. You have to know that Apple is only responsible for designing mobile phones, and they have never manufactured any of them but outsourced to FOXCONN owned by Terry Gou for their manufacture. Therefore, it can be said that Apple defeated Nokia with almost bare hands. In the era of the new industrial revolution, through collaboration and the sharing of inexhaustible network knowledge, the power of individual creativity will be magnified by many times. Sometimes a single Maker is enough to overthrow a traditional industrial empire, just as easily as taking something from a pocket.

Local Motors, an automobile manufacturing company, based on a network community platform mainly facing the personalized niche automobile market has successfully developed a car called Rally Fighter. The car sells for USD 75,000. The development only took 18 months and cost 3 million

US dollars, while General Motor's development of Ford needed 6 years and cost 6.5 billion US dollars. In 2011, the U.S. Defense Advanced Research Projects Agency (DARPA) collected inspiration from the public to design a replacement for their iconic military Hummer. In just 14 weeks, which is about 1/5 of the average manufacturing time in the automobile industry, Local Motors created their award-winning design into a working prototype, fully demonstrating the amazing strength and enthusiasm of the Makers.

There is another American Maker named Elon Musk. He is considered to be the role model prototype of the movie *Iron Man.* In 2002, he sold PayPal (the world's largest online payment platform, similar to China's Alipay) co-founded by himself and his partner to the world's largest online store eBay for 1.5 billion US dollars. After that, he founded a company, Tesla, and produced the first electric vehicle in the world that could accelerate from 0 to 100 km/h in 4 seconds, It was successfully mass-produced. In 2010, the Falcon 9 rocket launched by another company he founded, SpaceX, successfully parachuted the "Dragon capsule" back to earth. This is the first time in the history of the world that a private company launched a spacecraft into space and managed to receive it back. The entire aerospace industry was shaken. The standard launch cost of the Falcon 9 rocket is 54 million US dollars, compared to the cost for Aeronautics and Space Administration (NASA) which is 435 million US dollars. NASA subsequently announced that they would not build any new space shuttles in the U.S. after the existing ones retiring in 2011. Instead, they would entrust private companies like SpaceX to deliver supplies to the International Space Station. They signed a contract worth 1.6 billion US dollars with SpaceX. It can be said that in terms of rocket manufacturing and launch, Musk defeated NASA completely by himself. Moreover, Musk plans to invent a reusable rocket, hoping to realize the dream of human migration to Mars within the next 10 years.

It is the Makers' DIY abilities that keep opening up the sources of human innovation. Today, when 3D printing and 4D intelligent digital tools become the "nunchucks" in the hands of the new generation of Makers, the world will be changed. With the birth of the 4D printing concept and the development of its relevant technologies, Makers will, as always, receive more glory in the era of technology.

Section 6 | Combination of 4D Printing and Makers

The advantage of the traditional manufacturing industry is the standardization of assembly lines and its large-scale production, which is the most effective way for the development and profitability during a commodities shortage. However, in today's world where the production means and living materials are becoming saturated, the needs of consumers are turning towards cultural and spiritual thought.

In this process, new production means will be appreciated. For example, the emergence of new production means, such as intellectualization and 4D printing, on the one hand, conforms to the stage of our historical development. It creates the appearance of inevitable technologies under the development of science and technology. On the other hand, the changes in consumers' demands for commodities lead to the thinking and emergence of new technologies.

The emergence and popularity of the Makers is not a product of the era of industrialized mass production. In my opinion, it is more about thinking, innovation, and exploration of the niche and vertical fields, which are also the core of Makers' survival and development. In traditional industrial thinking, there is a proportional relationship between volume, cost, and profit. The reason why it cannot be broken is that the traditional industrialized manufacturing methods are usually based on the production process of molds and standard assembly lines.

That is to say, when the traditional industrialized manufacturing methods are used for small-batch or personalized product manufacturing, it is difficult for the input and output to achieve profitability. Therefore, industrialization is bound to rely on new production means and production methods to solve the problem of traditional manufacturing methods. This is also a core factor of why 3D printing is concerned in this movement and looks promising.

The impact of the revolution brought about by 3D printing in the manufacturing industry is far-reaching. It is satisfied with not only the production methods of "personal customization" but also the low-cost rapid manufacturing. Nevertheless, the impact brought by 4D printing will probably be even more thorough.

If 3D printed objects, such as furniture, take up more space in transportation and bring inconvenience during assembly, 4D printing can completely solve these problems. We only need to input the transformation memories into the printing material and models. Let's say a cup has water as its triggering medium, it can then automatically transform and adjust according to the amount of water stored in it. This means it won't be inconvenient for us if we wanted to drink a small amount of water from a big cup.

Of course, it is not limited to these products only. For Makers, 4D printing solves the problems of personalized production and manufacturing, meanwhile, it puts creativity into a pair of wings, which can self-transform and self-assemble. This pair of wings will be able to realize creativity, give more creative intelligence, and allow more creative ideas to happen within the products themselves in the coming era where everyone will be a Maker and everyone will be creative.

Section 7 | The Future of the Makers

On January 28, 2015, Premier of the State Council of the People's Republic of China, Li Keqiang, hosted an executive meeting of the State Council to determine policies and measures to support the development of "Mass Innovation Space" and build new platforms for entrepreneurship and innovation.

What has been pointed out at the conference were no longer slogans but some specific measures of great significance that were in line with the new situation where entrepreneurship and innovation were pushed by the era of a network. The construction of entrepreneurial service platforms, such as "Mass Innovation Space," forced everyone to stimulate their inspiration in innovation, nurturing all kinds of young innovative talents and innovative teams including college students to expand employment and create a new "engine" for economic development. This is of great significance.

From the current situation, Shenzhen, Beijing, and other places have begun to form the Maker industry clusters and the idea of a generation, product formation, and project funding support to the large-scale commercial production. In some places, Maker alliances and organizations appear one after another. But, from a national perspective, this is the first time the State Council has studied the above topics.

Building open platforms

The executive meeting of the State Council on January 28 pointed out that the next step was to vigorously develop the market-oriented, specialized, integrated, and networked "Mass Innovation Space" based on incubation models such as Makerspace and innovative factories to achieve the combination of innovation and entrepreneurship. As well as the combination of online and offline, incubation and investment, providing a low-cost, convenient, all-factor, open and integrated service platforms for the growth of small innovative enterprises and individual entrepreneurship.

It is encouraging that appropriate allowance to the rental, broadband network, public software, and others for "Mass Innovative Spaces" is being provided, as well as low-cost places for renting like the idle factories and other resources.

Meanwhile, it is necessary to improve the entrepreneurial investment and financing mechanism and activate the role of the government's Venture Capital Guiding Fund and finance and taxation policies. These would help to support scientific and technological small and medium-sized enterprises at seed and early-stage investment, and cultivate and developing angel investment.

To flourish entrepreneurship and innovation, we need to create a good ecological environment for these concepts, improve the entrepreneurial counseling and mentoring system, support and organize entrepreneurship training camps, competitions and activities, and cultivate Maker culture.

From the State Council's emphasis and specific measures proposed, relying on the Maker model as the entry point, a new round of national innovation and entrepreneurship craze will start very soon. In doing so, the demand for personalization and customization will inevitably explode and the transformation of traditional manufacturing methods and technical methods will also be promoted.

8

ETHICS

A seemingly ordinary strand was gently placed in the water, and a miracle happened over time. The strand slowly moved, rotated, changed, and went through twists and folds, to eventually form a regular tetrahedron. The amazing part was that there was no external equipment involved at all. At the TED conference held in Long Beach, California, the U.S., Skylar Tibbits, a lecturer from the Department of Architecture, MIT, demonstrated the fantastic 4D printing technology. When the movie-stunt effect, self-transformation, was truly presented in front of people, it was no surprise that it left the world with wonder and admiration.

When Skylar Tibbits, as the first person to reveal the topic of 4D printing technology that met the needs of humankind and even the planet in this unique way, issues related to intellectual property and laws in this field were raised. The same issues were entangling the development of 3D printing technology, but it was still a big source of concern for stakeholders in the relevant industries in the development of 4D printing technology.

SECTION 1 | LEGAL RISKS AND CRISES

A pair of jeans that self-repairs holes, a piece of furniture that stays "flat" in a vacuum and can automatically transform into a sofa when stimulated, water pipes that transform as the external environment changes, all these science-fiction-like ideas fascinate the people and raise doubts in them at the same time. Before we discuss the intellectual property and other related issues caused by 4D printing technology, let's take a look at the existing patent application status of 4D printing technology and its relevant materials.

When searching the keyword of "4D printing" in the Chinese patent database, there was no directly related patent documents found. When including and the keywords of "programmable materials," "transformation molecule" and "synthetic organisms" and biological membranes, no records were found related to the transformation molecule either. Yet, there were 48 patents found related to biological membranes. While searching in the foreign patent database, there were only two patents found related to the transformation molecule, and thousands of patents were found related to biological membranes. In conclusion, in the domestic and foreign patent databases, there are about ten patents for self-assembly related programmable materials, and the majority of them are used in the fields of DNA or nanostructures.

After this search online, experts in the industry expressed that the number of global patents found in the aspect of 4D printing-related self-assembling technology was almost 10,000. Regarding the molecular self-assembly technology, the number of patents found was close to 1,000 in the Chinese patent database, while the patents found exceeded 1,000 in the foreign patent database with the keyword of "molecular/biomolecular self-assembly." For memory metals, there were more than 300 patents found in the Chinese patent database with the keywords of "memory metals" or "shape memory," whereas, hundreds of patents were found in the foreign patent database. Based on the inference of the above, from the global patent layout, the current patent documents on 4D printing were still very few in number, and the enterprises involved in 4D printing are even less prevalent.

Presently, 4D printing is still in the early conceptual stages, and its relevant service markets are also in the infancy stages. So, there are very few people involved in this open ultra-frontier field, and the commercial atmosphere thus is yet to flourish to attract widespread attention from the public. However, as an emerging technology that will change the rules of its respective industries, its subsequent potential influence is unpredictable. Just like our personal computer, in only 10 years or so, it has fundamentally shaken our legal system.

In the foreseeable future, what new legal challenges, new forms of consumer safety and criminal activities will the 4D printing technology encounter?

Previously, making counterfeit currency was technically challenging. However, with the development of computers, the U.S. Secret Service highlighted in a public service announcement that the knowledge and skills required for making counterfeiter currency could be obtained from basic computer training, public education, and trial and error. Before 1995, less than one out of 100 of counterfeit currency was made using computers and laser printers, and just 5 years later, in 2000, nearly half of the counterfeit currency was designed online and then printed on high-end color printers. Computer design software, color laser printers, and toner technology have made it easier to fake counterfeit currency. The economic damage caused by counterfeit currency is self-evident. Through 4D printing, it can not only easily make counterfeit goods but also make guns, ammunition, other weapons, and even drugs without the country's permission. The risk of personal injuries to the people is chilling.

In 2012, the first small-scale ethical conflict in the field of 3D printing broke out on a file-sharing website called Thingiverse.com. Meanwhile, such conflicts will also appear in relevant fields after the emergence of 4D printing. According to the report, a user uploaded to this website a design file for printing a rifle part by using plastic as raw material on a consumer 3D printer. The particular rifle part was the only one that required a background check from the user in the gun manufacturing process, but the design file made it possible for people to own a rifle without a license. In other words, parts that are made through 3D printing technology can circumvent gun control laws. Even in the era of 4D printing, what the "finished product" or the printed "semi-finished product" will transform into under different medium conditions may be unknown even to the person who prints it. Unlike the appearance of a gun, it may look more like a shoe or a brush at first. This further increases the difficulty of the relevant authorities to monitor and control printed objects. Moreover, through the application of 4D printing technology, the manufacturing costs of guns will be significantly reduced. As the performance of the printed rifle was summarized by the netizens, the 3D printed plastic part was so powerful that it could fire 200 rounds. To complete this, there was no need for special equipment, just a long service life Stratasys printer could print out the part, and the raw material used was just ordinary commercial-grade resin costing USD30 only.

Other than firearms, the magical 3D and 4D printers will easily print out drugs, medicine, and many other unimaginable items. However, the more powerful the technology, the more far-reaching is its potential for abuse. The so-called DIY production of medicine breaks the relatively uniform and consistent formula for making medicine and compounds. It allows people to design and produce their medicine and powerful chemicals, which will no doubt be a nightmare for the regulatory authorities.

In the U.S., the "War on Drugs" has unfortunately failed. Prisons are full of non-violent criminals, and high taxes are being spent on arresting drug addicts rather than the lower-cost and more effective drug rehabilitation programs. The number of deaths from overdoses of prescription drugs in developed countries has soared. Imagine if someone makes a batch of chemicals that can change people's moods by using the printed reaction vessel and then spreads these products freely, what a catastrophe that will be!

Section 2 | The Lost Intellectual Property

The sudden advent of "4D printing" while "3D printing" is still becoming a hot topic today has amazed many people. It is considered as the "signs of the third technological revolution." Even U.S. President Barack Obama once mentioned these terms in his State of the Union address!

The 3D printing technology is incredible. It allows manufacturers requiring complicated craftsmanship under the traditional technical conditions to complete it by a simple click of a mouse, and a variety of crafts, toys, clothing, and violins could be easily printed out. 4D printing is even more fantastic, making the "Transformers" in science-fiction films and the "Jingu Bang" in the myths a reality. The printed objects could transform into anything you wish.

Obviously, the impact of 4D printing on the future of the world will be huge. Among all these impacts, the protection of intellectual property has firstly entered people's field of vision and has become one of the people's focus and concern.

Looking back over the past two years, there have been many disputes over the intellectual property during the rise of 3D printing. A British game company issued a cease-and-desist "order" to a 3D printing vendor for printing physical models of the character from the company's popular desktop game "Warhammer." A Netherlands designer sent an "offline notice" to Thingiverse, which is an online database of design codes for 3D printers that allows anyone to download and share designers' designs for free. The controversial Swedish file-sharing website, The Pirate Bay, which has been the target of lawsuits around the world, recently announced that it was embarking on a new service to share 3D printing designs. The American cable network HBO sent Fernando Sosa a "cease-desist letter" asking him to stop selling 3D printed iPhone docks mimicking the Iron Throne in the HBO series "Game of Thrones."

It is apparent that the legal protection of intellectual property of 3D printing is still an issue that needs to be considered and urgently resolved. Yet, for the intellectual property of the coming 4D printing, it's more like a lost lamb – where will it be led to?

We will shudder if we follow 3D printing's thinking on intellectual property issues because 4D printed items could change shape at any time according to the changes in the set time dimension and stimuli. Let's assume that a designer 4D prints a ring, which is just a representation of a semi-finished product, and ultimately it may be a rose, a gun, or even drugs. Such changes in 4D printing are not just a dream or hypothesis. Then, without a setting in time, people won't be able to identify what the printed item will ultimately be, thereby they won't be able to evaluate the ownership of the intellectual property. However, even though the time will be preset, it will be meaningless for 4D printing as people could avoid the preset time while designing.

In the face of such embarrassment, can we boldly and tolerantly assume that 4D printing is more like children playing with building blocks? Whatever gets built is what it is? If this is the case, who can say that what a person has built must have a patent and intellectual property protection, while others cannot build the same or similar things? In fact, given this scenario, the issue of patent and intellectual property will be shifted to building blocks instead of how they are built. In the same way, the intellectual property of 4D printing, can create a new situation by shifting these concerns. It's worth waiting to see what will happen.

Section 3 | Challenges to Intellectual Property Protection

Intellectual property protection is becoming more and more important in the period of the rapid development of science and technology. Within the current background of mass entrepreneurship and mass innovation, only a perfect system of intellectual property protection can bring lasting power to innovation and economic development. With the conditions of traditional manufacturing technology, including personalized customization based on Industry 4.0 technology, although the production cost of personalized customization has decreased compared to traditional manufacturing technology, the production cost of personalized customization for the individual user is still high compared with 4D printing technology.

This means that with 4D printing technology, production costs will be greatly reduced, especially with the increasing maturity and popularization of desktop printers. This will fundamentally hinder or shake the interests of traditional manufacturers. It is foreseeable that in the era of 4D printing, few people will be willing to pay high prices for branded products, but rather buy low-cost raw materials and print products at home with their 4D printers. As we can see, 4D printing poses new challenges to the existing fair use system of intellectual property, as well as the monitoring and evidence collection systems. In particular, the difficulty of intellectual property protection caused by the 4D printing's capabilities of self-transformation and self-assembly triggered by adding the time dimension will be far beyond those of 3D printing.

Currently, we still lack the related intellectual property protection systems for 3D printing technology. Compared with 4D printing technology, intellectual property protection for 3D printing is relatively simple and it focuses more on how we could solve the infringement behaviors of "imitation and copies." However, 4D printing technology creates a further subversion and transformation from 3D printing. The addition of the time dimension makes the printed objects no longer a simple copy but it integrates factors of self-transformation and self-assembly, which pose a challenge to traditional intellectual property protection.

The current problem is that the existing laws of intellectual property protection, blank on both, 3D printing and 4D printing. This is not only a problem in China but also a common challenge faced by countries around the world. Now, the problems faced by 4D printing are that as technology advances, various requirements of intellectual property protection will be inevitably put forward, and there will therefore be mainly three challenges. These are described below.

The product form of public transactions. Due to the self-transformation and self-assembly capabilities of the 4D printed products, the intellectual property protection of commodity transactions is more complicated. One example is a home appliance. The 4D printed home appliance is a copy of a certain brand, and while going through the evidence collection process conducted by the regulatory authority, it can self-transform and self-assemble into another form which is non-infringing. Let's say a refrigerator is copied from product A. In the course of a display or an evidence collection, the refrigerator could self-transform and self-assemble into a form with different size and shape that has been preset in the changeable design model. This makes it somewhat more difficult to obtain evidence for intellectual property protection.

The grey edge of customization transactions. Just like some business models and services were derived from 3D printing, there will be some professional printing platforms that will provide services to customers derived from 4D printing, as well. But, the forms presented by the platforms will be more diverse and varied, perhaps they will follow a crowdfunding model, or belong to a few people with a common interest serving a community or a group. With the help of software synthesis technology of 4D printing, when users see a product they like anytime and anywhere, they could take a photo of the product by using a cellphone or a wearable device and send the photo to the selected customization platform. Then, the platform providing a 4D printing service will synthesize the photo with the aid of software and print out the corresponding product for the user. Such service derived from customization will push intellectual property protection to the grey edge.

Copyright challenges for personal printing. When desktop 4D printers enter households, especially for some smaller products, such as pots and pans, teacups, home decorations, among others, users could then complete the printing at home depending on their preferences. When users surf the internet and find a product they like or a creative design they are interested in, they could simply download the image to the corresponding 4D printing software for 3D synthesis and then send it to their 4D printer to print out the product. Not only that, but users could also print a variety of different designs into one and trigger the printed object to transform to fulfill their needs when needed. Thus, we can imagine that difficulty of evidence collection of infringement could occur in the process.

Due to the misalignment of law updates and the development of science and technology, the loopholes caused in the legal perspective of intellectual property protection have been presented. Therefore, the questions of how we can adapt the requirements of new technology development, find a balance between new technologies and innovations and establish a protection system for 4D printing's intellectual property will soon be the possible common concerns facing the industry experts and regulatory authorities. As far as I am concerned, we should explore and establish relevant laws and regulations without delay to better promote the healthy development of the industry and minimize the negative impact brought about by the new technology revolution.

Section 4 | Copyright

Intellectual property is a large legal concept that includes copyright, trademark rights, and patent rights, among which, whether we are talking about 4D printing or 3D printing, copyright is the most closely concerned concept.

Copyright is primarily applicable to creative works. It gives the original copyright owner exclusive rights within a certain period of time to arrange their work (the specific period varies from country to country, but it is typically the life of the copyright owner plus 70 years after their death).

If others or other organizations want to reproduce, modify, sell, rent, or publicly present the work, they must obtain permission to do so from the copyright owner.

The key to copyright is the price of the artwork and not its usage or practicability. Michael Weinberg, a lawyer, and a 3D printing expert, once explained that "making things" itself cannot be granted copyright. The scope of practicality is wider. For example, clothes are useful, and you can have the copyright of clothes' design but not for the concept of tailoring.

Industry experts predict that the most challenging issue in copyright disputes in the future would be the recognition of copyright of "derivative works." According to Copyright Law, a derivative work is an expressive creation based on the original work. Translation is a derivative work, an imitation of a popular song is also a derivative work. If you edit a design file of existing work and turn it into a new one, under what circumstances are you creating a derivative work? Under what circumstances are you creating a brand new work (a brand new work means you are the copyright owner)? Or under what circumstances are you deliberately taking the fruits of others' labor?

4D printing can be regarded as a kind of "derivative work" in some ways. The concept of copying "works of applied art," and illegal copying are exactly what copyright must prohibit, particularly for the copyright protection of "product appearance and structure" involved in 4D printing. The current Copyright Law in China is not sufficient to protect the copyright of "product appearance and structure." Those few "product appearance and structures" with artistic value can be protected as artworks, while it is difficult to protect most ordinary "product appearance and structures" under the Copyright Law. The Copyright Law of China is facing a third revision, which draws on the relevant provisions of the Berne Convention. It is adding provisions on copyright protection for "works of applied art," and formally including the "product appearance and structure" into copyright protection.

In the process of 4D printing, if the "copying" is done without the author's authorization, it is considered an infringement. However, while judging the infringement of 4D printing, the method of printing also needs to be looked at. So far, no matter if we are discussing 4D printing (whereby the semi-finished product produced by the printer with the time dimension transforms into its final form triggered by a specific medium) or 3D printing that directly prints product into a form, both processes involve the following three methods. First, the method of printing from three-dimensional to three-dimensional object, that is, the printing of a three-dimensional model from the computer to a physical object. Second, from text to three-dimensional, that is, entering textual descriptions on the computer, such as rectangle, height 18 cm, width 20 cm, red in color, etc., to print out a semi-finished object or the corresponding object. Third, printing from a flat to a three-dimensional model, that is, printing the flat image on the computer through the printing program into a three-dimensional object which will transform into its final shape by triggering the medium.

The first method of three-dimensional to three-dimensional printing is a typical copying behavior. For flat to flat or three-dimensional to three-dimensional printing, both methods belong to the typical copying when it comes to the Copyright Law. Even if it is a reduction printing, and enlarging printing, or other methods of changing the proportions, it won't affect the fact that it would be considered copying. Therefore, copying without the author's permission will constitute an infringement.

The second method of text to three-dimensional printing usually is not considered as copying under the Copyright Law. The Copyright Law protects "expression," and text and three-dimensional printing are two different forms of expression, so it doesn't involve copying. For the same reason, this printing method for 4D printing and 3D printing generally won't be affected by the infringement.

Is 4D printing from a flat model to a three-dimensional model considered copying? China's Copyright Law is silent on this topic, but it is quite controversial in practice. In the "Fudan Kaiyuan case" in 2006, the defendant converted the form of the Chinese zodiac from flat cartoon images to three-dimensional money-boxes without legal authorization, and the court held that this was an infringement of the plaintiff's copyright. However, in the "Motorola copyright case," the court held that Motorola's act of producing printing circuit boards following their design drawings was an act of industrial production and thus did not belong to the act of copying under the Copyright Law. The court's judgment on whether the same "flat to three-dimensional" method constituted copying was completely different. When we refer to the provisions of the Berne Convention on "copying," which includes copying in any form and any way, such open-ended phrasing raises a higher standard for copyright protection. More importantly, in the era of 3D and 4D printing, such "copying" methods will inevitably multiply. So, it is necessary to clearly state the "copying" methods in legislation to protect the legitimate rights and interests of copyright owners.

Section 5 | Trademark Rights

Much like copyright, trademarks are also considered intellectual property. People usually use "brand names" or "logos" to define a trademark. The information it conveys is that it is a registered mark or "trade dress" that shows consumers that the product is produced or provided by a specific manufacturer. The original purpose of trademarks was to protect the interests of consumers. Gradually, trademarks have since become a marketing tool, and those highly recognizable trademarks can be worth billions of US dollars.

In the development of 4D printing, it is likely, that trademark infringement will also be involved. For example, users may print out the trademark together with the product, such as NIKE shoes. A printed pair of NIKE shoes will usually carry the NIKE trademark, and if this printing does not obtain authorization from NIKE, it will very likely constitute an infringement. Other than that, for the rest of the printing methods, if one is simply printing the product itself, that is, only printing out NIKE shoes without the NIKE logo, generally, this should not involve trademark infringement but may result in other intellectual property infringement.

Much the same, 4D printing technology also opens up the door of convenience for the production and manufacturing of trademarks. Although this is a relatively independent printing behavior, even if the user only prints out the NIKE trademark without printing the shoes, this printing may also lead to trademark infringement. This is because according to China's Trademark Law, the act of manufacturing and selling a trademark without authorization from the trademark owner would be considered a trademark infringement.

While 4D printing, users can also print according to their ideas, such as by following a printing method like only printing NIKE shoes (with no trademark), but at the same time, later printing the PUMA trademark on the shoes. This means that the NIKE shoes will carry a PUMA trademark. Would we consider this an infringement? Whose trademark right does it infringe? First of all, if the user doesn't have authorization from PUMA, the unauthorized use of the trademark may constitute its infringement. Then, what about NIKE? Does such a printing method constitute a trademark infringement for NIKE? In China's Trademark Law, there is a provision called "reverse infringement," which means if you replace the NIKE trademark and put the products on the market without permission, it still belongs to trademark infringement. So, if a user prints out products by using the above method and puts them on the market, their behavior infringes the NIKE trademark as well.

SECTION 6 | PATENT

The word "patent" is derived from the Latin "patere," which means "to lay open." According to Wikipedia, a patent is "a form of intellectual property that gives its owner the legal right to exclude others from making, using, or selling an invention for a limited period of years in exchange for publishing an enabling public disclosure of the invention. Like any other property right, it may be sold, licensed, mortgaged, assigned or transferred, given away, or simply abandoned."

The core technology of 3D printing, "laser sintering," has a patent. In January 2014, the patent protection period expired, which allowed the 3D printing patent to enter the public domain, largely promoting the rapid advancement of 3D printing. With the advent of a large number of 3D printers, the manufacturing revolution is now ready to take off, and in this process, the issue of patent infringement will undoubtedly attract more attention.

According to China's Patent Law, the types of patents can be roughly divided into three categories: inventions, utility models, and designs. Invention and utility model patents focus on the internal structure and innovation of a product, while the design patent focuses more on the appearance and color of a product. Unlike the existing flat printing, either 3D printing or the further development of the new 4D printing involves both the appearance and the internal structure of a product. Hence, the three categories of patents are closely related to each other, which undoubtedly requires a user to be familiar with the patent protection of the product they want to print in advance in order to prevent infringement of others' patents.

The development of 3D printing and 4D printing may also promote the commercial use of expired patents. According to the provisions of China's Patent Law, patents have a protection period. The protection period of invention patents is 20 years, and it is 10 years for utility model patents and design patents. If the protection period is exceeded, the patent will enter the public domain and everyone could use it for free. Many of the expired patents are full of creativity in appearance and structure, but they are limited by technical limitations and are difficult to realize. Therefore, they are left unused. An American lawyer has discovered a lot of interesting and useful designs out of the expired patents and set up a special section on the Internet for users to download the expired patented 3D design drawings for their own printing. This will undoubtedly be used in the field of 4D printing.

SECTION 7 | "FAIR USE"

When discussing 4D printing and intellectual property infringement, we must also look at the issue of "fair use." So, what kind of behaviors is called "fair use?" As regulated by China's relevant laws, "fair use" in the Copyright Law refers to the use of a work in a certain way without the consent of the copyright owner or without the payment of remuneration to the copyright owner. In the intellectual property system, acts of copying a small amount of published work for personal use only, such as for self-study, research, appreciation, use in the classroom for teaching or teaching research, among others, are regarded as "fair use." For these instances, the permission of the copyright owner is not necessary and the use is excluded from the scope of infringement.

What if a user simply 4D prints personal consumer goods and does not use them commercially, does their behavior constitute infringement? Under the existing Intellectual Property Law, such behavior is generally regarded as "fair use" rather than infringement.

If this is the case, we can imagine that in the future society, few people will spend a lot buying branded products. Instead, people will be more willing to spend less on buying low-cost raw materials to print out the needed products at home. The result of "fair use" by many people will lead to the bankruptcy of merchants. With no surprise, 4D printing also challenges the current fair use system.

Under the conditions of traditional technology, the production cost of the individual user is relatively high. Therefore, production for "research, appreciation or personal use" is extremely limited in scale and does not fundamentally hinder the interests of merchants, so much so that there is no problem with the behavior being considered fair use. But, under the conditions of 4D printing, the production cost will be greatly reduced, which will make the idea of "fair use" fundamentally hinder the interests of merchants. By then, revisions of "fair use" clauses will be needed, and a contest between the public and patent owners will be inevitable.

Even now, the adoption of technical means to prevent "fair use" from damaging the interests of merchants has been put on the agenda. The United States Patent and Trademark Office recently introduced a "production control system" for copyright protection of 3D printing. Under the system, before any 3D printing-related device performs a printing task, it must compare the printed model with the data in the system database. If there is a large proportion of matching, the 3D printing task cannot be performed. Both the Digital Millennium Copyright Act of the U.S. and the Copyright Law of China support copyright protection through technical protection measures, and behaviors that breach technological protection measures are regarded as illegal acts.

Section 8 | Protection of Consumers

Along with the maturity and popularity of 4D printing technology, when more and more people use 4D printing to produce the items they need, potential risks will continue to grow and spread while they enjoy the personalization and convenience. These risks go beyond the traditional legal framework of consumer safety. Under the existing legal mechanisms, firstly, no one can ensure that consumers will not be able to buy products that hurt them. Secondly, once harm occurs, there is no way to attribute responsibilities between the parties involved.

According to China's Tort Law, if a party can prove that they have made reasonable efforts to prevent and minimize the risk of product failure in the context of fair use, the party's responsibility ends. A simple example is tire manufacturers. If the tires they produced exploded every time the speed exceeded 120 kilometers per hour, they will be held liable even if it is illegal to speed at the time. This is because speeding is still common in everyday life. However, if the buyer of the tire drives into the roadside and the tire explodes, the tire manufacturer will not be held liable.

Standards help to set clear boundaries of responsibility. In the early days of steam engines, boilers often exploded, frequently causing damages. In response to this situation, insurance companies have set explicit production standards to divide responsibilities. Finally, a series of standards have been established, such as standards of material thickness, safety margins, and pressure relief valves, to ensure that filters could withstand the lowest steam pressure requirements. Boilers that do not meet these standards will not be eligible for insurance compensation.

Such a legal mechanism with a clear allocation of authorities and responsibilities is also worth referencing during the age of 4D printing. In contrast to the current situation when 3D printing is popular, an inevitable scenario can be that a car hobbyist could be killed in a car crash due to the disengagement of the purchased 3D-printed steering wheel while driving at high speed. When the defense lawyer is defending the victim's family, they could be confused as to where they could attribute the responsibility. Should be the person who created the wrong design file? The person who 3D printed and sold the part? The websites which advertised the part? Or the manufacturer who installed the part on the car, the manufacturer of the 3D printer, or the raw materials supplier?

In the absence of clear and standard boundaries of responsibility, no right of recourse would be imposed on manufacturers of 3D and 4D printed products and as such, they could irresponsibly seek to maximize profits. Such a situation may be familiar to everyone. It is similar to the user agreement we have once read, whereby we were only allowed to install new software after we accepted its terms. Most software will be sold with such a statement, "VENDOR DOES NOT WARRANT THAT THE FUNCTIONS CONTAINED IN THE SOFTWARE WILL MEET YOUR REQUIREMENTS." The same thing happens during Taobao's transactions, which is flooded with counterfeit goods.

Conversely, when all behaviors have a clear and standardized system that defines the boundaries of responsibility, perhaps things will become more harmonious. In the face of standardization under the requirements of laws and regulations, 4D printer manufacturers will seek to certify printers. At the same time, manufacturers of 4D printing raw materials will also try to meet or exceed the minimum material performance standards for manufacture and production. Meanwhile, 4D printing source designers will also constantly check and improve the accuracy of the digital models.

Section 9 | Low Energy Consumption for the Earth

In light of the world energy and resource crisis, the destruction and waste caused by traditional manufacturing remain high. The advent of 3D and 4D printers at this time is being entrusted with a mission with critical and difficult timing.

Recently, at Die & Mould China exhibition held in Shanghai, a technician at the booth of international 3D printing giant, Stratasys, said, "There is almost no waste of materials in this production method. It is not only cost-saving and brings environmental benefits but also enables localized production and reduces the need for transportation. Customers only pay for the design and then print out the required product locally, thereby reducing the need for long-distance transportation and greenhouse gas emissions."

Indeed, the manufacturing of products based on 3D and 4D printing technology will reduce the waste of materials and carbon emissions in many ways.

The performance of 4D printing, in particular, is even more outstanding. It only uses the materials that make up the product and does not need any extra materials, but it also maximizes the utilization of raw materials in the manufacturing process. As demonstrated by Stratasys staff, it is possible to send the three-dimensional image data through a computer to a 3D printer to directly "print" the three-dimensional image into a physical object, as per the function of 3D printers. It works in much the same way as an ordinary printer that is connected to a computer, with liquid or powder printing materials, which then "printing materials" layer by layer through computer control to eventually turn the blueprint io the computer into a physical object. The entire production process is a bit like building blocks, materials are added layer by layer until the final shape is complete.

According to Wang Xiang Gen, General Manager of Stratasys-China, the 3D printers they currently produce are mainly for R&D and rapid manufacturing purposes. "Because 3D printing is suitable for various small batches, and personal-customized products, for a long time, 3D printers have been mainly used in R&D. However, with the maturing and cost-reducing technology, more and more fields are starting to use 3D printing to produce end-point products." Today, Stratasys has been providing 3D printing solutions for a wide range of industries to optimize traditional manufacturing processes. The range of customers covers aerospace, automobile, medical, and dental to education spheres, consumer goods, consumer electronics, and other fields. At the exhibition, the display of Objet500Connex3 by Stratasys, which was the only color printer and multi-material 3D printer in the world at the time could increase the variety of printing materials available for 3D printers from more than 100 to more than 600. This shocked the industry. Wang Xiang Gen predicted that as the number of materials available for 3D printers increased, there would be more fields that 3D printing technology could be directly applied to for rapid manufacturing.

Joshua Pierce, a Michigan professor, has published a report stating that the direct application of 3D printing in the manufacturing field is not only cheaper but also more environmentally friendly. He found that the energy used of an object produced by a basic 3D printer was 41% to 64% less than the one produced by traditional technology. It also saved on the use of raw materials. For example, some molds produced by traditional technology were usually solid, while 3D printing could produce partially or even completely hollow molds.

More importantly, the localized manufacturing of 4D printing will overall reduce the amount of products' transportation. Since the beginning of the industrial revolution, people have been increasingly relying on complex production technologies, and the places of manufacture were far from where people lived. Since human beings became obsessed with globalization in the 1980s, the places where goods were produced and sold could be tens or even hundreds of miles away from each other. Now that people can buy goods thousands of miles away, this has become so common that a significant portion of a product's value goes to transportation. Perhaps, economists and corporate accountants who are more concerned about economic benefits do not add the petroleum supply reduction and climate change to their theories or accounts. Yet, it is undeniable that these issues have become a heavy burden to social development and even the global environment. With the popularity of 3D and 4D printing, people will be able to manufacture many things locally and sometimes even directly produce and assemble products at home. As a result, a wide range of local businesses supported by 3D and 4D printing equipment will be more capable of meeting the needs of local customers. This means that the current demand for importing a large number of goods from overseas will also gradually disappear. The only thing left that will need to be transported may be raw materials, which take up less space during transportation.

Section 10 | Product Repair and Material Recycling

The use of 4D printing will increase the service life of the designed and manufactured products while reducing the loss of raw materials in the manufacturing process. The products' storage of digital objects and localized manufacturing would not only minimize the problems in the product design process but also highly improve the convenience and possibility of product repair.

The concept of 4D printing, based on the 3D printing's equipment and the storage of digital objects, will correct and improve the problems in product design. The iPhone charging dock, called an Elevation Dock, is a typical case. Elevation Dock is a successful innovation that raised funds through the Kickstarter website. At the time, an Indian designer designed a perfect iPhone charging dock, which successfully raised 1.4 million US dollars, and then he delivered the charging dock to the sponsors. But it was the late delivery that created a very serious problem, which was that when the iPhone 5 was officially launched, it was using the new Lightning connector.

A London software engineer, Mike Hellers, explained his miserable situation, "I actually received mine the day before I received the iPhone 5. Couldn't have been worse, really, as it meant that I used the dock with the 30-pin connector for exactly one day before it became a useless paperweight." However, as an enthusiast of 3D printers, Mike Hellers was not as pessimistic as other sponsors. Instead, he took this as a challenge. He opened Tinkercad, a web-based 3D model design software on his computer, turned on his MakerBot 3D printer, and then began to design the adapter. It didn't take him long to post his designed adapter on the website Thingiverse. He then interacted with people who didn't have a 3D printer on Shapeway, and more than 12,521 users downloaded the adaptor and were able to continue using the existing charging dock.

Matt Haughey, the founder of Metafilter, saw the adapter design on Thingiverse and thought there might be room for improvement. So, he and his friend Michael Buffington printed out some samples and found that the design couldn't hold the wires flat, which made the adapter easily fall out. Their improvement managed to solve the problem. They printed out a sample of the new design for testing and then shared it on Thingiverse. Haughey said, "It only took a couple of hours of tinkering through two existing designs to meld the best parts of each into something that worked better than previous designs."

A user of a 3D printer came up with a solution within a few hours, but it took the Elevation lab more than one month to work out an official solution. Actually, in hardware manufacturing, five weeks to turn things around is usually a pretty good result. But compared to the users of 3D printers, waiting five weeks is too long. "Any industry that involves physical objects and small parts is ripe for 3D printing." Haughey firmly believed that high-definition 3D printers would open up a whole new industry and play a critical role in repairing the components of used products.

In the past half-century, people in developed countries have been constantly forced to replace their commodities. Especially for the older generation, it may take quite long for them to save enough money to buy daily necessities. However, most people from the younger generations could choose to buy commodities via loans, and if the commodities do not work before they are paid off, that would be unacceptable for them from both, a personal or environmental point of view. Besides, the discontinuation of the components' production repairs the "can" to the "impossible."

Furthermore, the current market is full of products with similar functions mainly because only a few products contain replaceable components. We can blame poor product design, and also the fact that most manufacturers are expecting us to keep buying new products rather than repairing them for us. Therefore, even if a product has excellent repair service, it is not cost-effective for the manufacturer to just replace the components.

If the local 3D and 4D printing equipment are widely distributed, they will change the status quo of product repair. Most products' components could be stored digitally and then manufactured as needed. Even if some components cannot be found on the Internet, a smart person (or a future repair shop) will be able to design the component, or scan the damaged part and repair it digitally on the computer, and then print out the component accordingly. As Jay Leno recently pointed out at Popular Mechanics, when he wanted to buy a component for a 100-year-old car, he could simply scan the component and have it produced through 3D printing or 4D printing.

All these behaviors, which are "superpowers" to the traditional manufacturing model, will gradually become many people's instinctual reactions in the current era of 3D printing and the coming era of 4D printing. It is also because of this, that when our watch stops working, our mobile phone is broken or our laptop is too old to run, we could print a new component for replacement to make sure we do not "abandon" it due to the lack of components.

Not only that but when we don't want some items anymore, we can recycle them through 4D printing, restoring their raw materials for rebuilding to maximize the use efficiency of raw materials as much as possible.

A young inventor, Tyler McNaney, created a "Filabot," which is a "plastic recycling machine" that grinds various plastics, such as milk jugs, soda bottles, all kinds of discarded plastics, and old 3D-printed objects, and then makes them into new usable printing filament materials. This will allow 3D and 4D printing to possess really strong sustainability qualities.

Section 11 | The World's Resources Optimized by 4D Printing

In January 2012, a seed report from the United Nations Conference on Sustainable Development summarizing our current situation stated that the challenges we are facing were severe. Yet, when we look at the old problems from a new perspective, perhaps we will be able to find many new opportunities, which include technologies that could pull us back from the brink of the earth's extinction. At the same time, the report also told us that the ecological pace of human development has been far beyond the earth's capacity, and the earth's limited resource supply has begun to seriously shrink. To avoid environmental catastrophes and maintain normal living conditions, we must do more with fewer resources over the next few decades.

The most pressing resource challenge presently is an "oil peak." It refers to the point in time when global oil production reaches its maximum, after which the oil demand will begin to exceed supply. Although the exact timing of an oil peak is a matter of debate, there is no doubt about its certainty. The most pessimistic prediction has already been made – the year 2015. To obtain the most accurate date, in 2009, the UK Energy Research Center analyzed 500 studies, as well as some major industry databases and 14 global oil supply forecasts. It concluded that, "a peak in conventional oil production before 2030 appears likely and there is a significant risk of a peak before 2020." Because humans are so dependent on oil, the world will be in serious trouble when it reaches its oil peak. After all, our transportation uses 50% of the oil, machinery, and power stations need oil for fuel and the large-scale manufacturing industry also needs oil. The use of oil includes plastics, inorganic pesticides, pharmaceuticals, pigments, fiber fabrics (such as nylon and polyester), other petrochemicals, and even the asphalt on the road is produced using oil.

While worrying about the oil crisis, we have to face the reality that oil is not the only scarce resource that will peak in the next few decades. For example, the United Nations issued a warning many years ago that freshwater supply will also be threatened, and a"water peak" will reach many parts of the world in the foreseeable future. Throughout the 20th century, human consumption of freshwater increased by 9%, and now 50% of it is being used by humans. Such usage has far exceeded the range that many regions can afford, leading to a sharp drop in groundwater levels in these regions, including China, India, the Middle East, and the southern U.S. The United Nations predicted, "By 2025, 1.8 billion people will be living in countries or regions with absolute water scarcity, and two-thirds of the world's population could be living under water-stressed conditions."

As a direct consequence of the shortage of oil and water supplies, global food shortages are likely to become increasingly widespread and severe. In many developed countries, every calorie of food on the table consumes 10 calories of oil, including the use of pesticide production, agricultural machinery, and transportation, refining, packaging of food, and other factors. The reduction in the amount of oil available (increase in the prices of machinery) will reduce food production. Besides, 30% of the world's agricultural land capacity is now declining due to soil erosion and over-intensive farming. Also, global fish stocks are falling. The 2010 Census of

Marine Life report stated that if no measures were taken, global commercial fishing would completely collapse before 2050.

Many scientists believe that food production is also affected by climate change. Now, rising temperatures have been linked to water shortages and soil erosion, which has also led to a decline in food production.

9

THE FUTURE (PART 1)

The Next Wave: 4D Printing – Programming the Material World, a research report by the Atlantic Council of the United States, discussed new disruptive technology, 4D printing. This was the first time that the scientific community formally discussed and deliberated the issues of 4D technology and defined 4D printing technology as the next wave, representing the human society moving towards a bright future. There is no doubt that 4D printing will transform the traditional manufacturing industry and reshape the future commerce with its advantages such as high energy efficiency, low cost, and personalization. In the development of 4D printing technology, the scientific community will take the lead in solving constraints like equipment, materials, software development, and other aspects, to promote the popularization and application of 4D printing technology.

Section 1 | The Ten Advantages of 4D Printing

The transformation that will be caused by the emergence of 4D printing technology will be far greater than that of 3D printing technology. Its impact on business forms and the transformation it will bring to the manufacturing industry will be far-reaching. For the manufacturing industry, 4D printing technology will bring at least the following ten advantages.

Advantage 1: Significantly reduced manufacturing cost

With the current modern manufacturing technologies, whether "Industry 4.0" in Germany, "Industrial Internet" in the U.S., or "Made in China 2025" in China, the cost of either manufacture or assembly of complex and personally customized products is relatively high. The cost also rises in proportion to the complexity of the components used.

However, when it comes to product manufacturing supported by 4D printing technology, the complexity of the components, and the structure of the product becomes less important. Because of the integrated printing of different components of the whole product, together with the ability of self-assembly, the assembly cost will be minimized and the product manufacturing cost will be significantly reduced.

Advantage 2: Unchanged cost of personalized customization

Under the present manufacturing technology and environment, the cost of small-batch customization is still high. Even if we fully implement the technologies of German's "Industry 4.0" or China's "Made in China 2025" tech today, we still cannot solve the problem of the high cost of small-batch production.

But with the intervention of 4D printing technology, the cost of personal customization will be on par with that of traditional mass manufacturing, and it may be even cheaper. This is because the considered complex structures in traditional manufacturing will be simplified with 4D printing. Particularly the cost of 4D printing will not fluctuate by the complexity of components or the labor cost of assembly.

Advantage 3: The released manual assembly cost

A key difference between 4D printed products and 3D printed products is that some 3D printed components need the secondary assembly to be completed. 4D printed products do not need to be assembled by manufacturers or users. Manufacturers only need to print out the desired designed product by the user and then deliver the printed product to the designated location.

The user can directly trigger the medium according to their demand to enable the product to self-assemble, which therefore replaces the method that currently relies on manual assembly and manual disassembly. The labor cost will therefore be greatly reduced by the change of production methods brought about by 4D printing technology.

Advantage 4: Zero-inventory production methods

For manufacturing enterprises, all inventories are assets that will fluctuate with the changing market. Usually, except for special products such as alcohol, the longer the products are kept, the lower the value of the products becomes. For today's business forms, inventories are also a core part of an enterprise's capital cost. Once sales drop, it will directly lead to a decrease in capital turnover, thereby affect the enterprise's profits.

Production through 4D printing will effectively alleviate this problem. Manufacturers could provide product designs based on users' ideas at any time and then print out the corresponding products. In this way, the traditional method of inventory sales will be replaced by the method that can achieve immediate buy-as-made and make-as-sold models.

Advantage 5: Enlarged creative space

To many designers, the most painful point is that creativity is beautiful, but the reality is cruel. A lot of creativity is constrained by various traditional processing and manufacturing techniques. A design that looks like a work of art is often changed into an entirely different looking product.

Therefore, with no exaggeration, as long as designers can come up with creative ideas conforming to physical principles, 4D printing technology will be able to turn them into reality without compromise. This will allow the value of the creative design to fully bloom. So, for 4D printing technology, different creative ideas at the technical level will have the same degree of difficulty.

Advantage 6: Lowered manufacturing expertise

Today's simple or complex manufacturing has put forward requirements of expertise and proficiency to manufacturers to a certain extent, especially in some manufacturing fields requiring high expertise. Besides, it has also placed considerable demands on the workers' expertise and proficiency.

Why do many products manufactured in China, particularly high-end equipment, always lag behind some European and American products? It is because the skill level of our manufactures is behind theirs, and the cultivation of technical personnel requires years of training and accumulation. It not only takes a lot of money to cultivate technical personnel but also bears the risk of brain drain. However, the application of 4D printing technology will reduce the specialized skills required for manufacturing complex components and will help us take on the manufacture of highly sophisticated crafts. Thus, the expertise threshold of manufacturing and the risk of brain drain will be effectively lowered.

Advantage 7: Simplified manufacturing process

Products nowadays are often assembled using multiple components, and different components require different equipment to manufacture them. For manufacturers, this not only needs a large space but also requires large capital investment in building equipment.

However, 4D printing technology is different. Just with a printer, and based on different materials and the different shapes of the components set by the user, it can directly print out the components or the whole product. In particular for the whole product, as long as the various assembled and transformed models are embedded in the initial printing model, the printed product can self-alter to completion, even with the most complex assembly method.

Advantage 8: Defective rates are a thing of the past

The control of products' defective rate by traditional manufacturing enterprises is a vital indicator of the enterprises' economic benefits. In the era of 4D printing, the term defective rate may disappear as part of history of the manufacturing process, and the key to determining whether the products meet the requirements of users will shift to the process of design.

From a theoretical perspective, as long as there are no problems with the printer, printing materials, and the designed model, the printed products won't have any defects. In the future, the critical factor to determine whether a product is of good quality will no longer rely on manufacturing, but the design of designers. Therefore, the bad or good quality of 4D printed products will result through design, rather than manufacture.

Advantage 9: Infinite combinations of materials

Combining different raw materials into a single product is difficult for today's manufacturing techniques because traditional manufacturing machines cannot easily combine multiple raw materials during the cutting or mold-forming process. Even though the molding multi-material hybrid injection that has been applied in some fields, its cost and defective rates are high. The advent of 4D printing technology, however, has made it possible to mix and print multiple materials on one device, no matter if the materials are plastic, metal, or other synthetic. This means that in the future, a complete and transformable car could be produced just by printing.

Advantage 10: Perfect consistency in mass production

Although modern manufacturing techniques guarantee the consistency of production to a certain extent with the help of molds, there are still a variety of differences between products. It is difficult to ensure 100% consistency and accuracy. However, this problem will be solved with 4D printing, where mass production is as simple as copying a digital file and as consistent as a digital copy. 4D printing technology will also be able to ensure the consistency of production in both large and small batches.

These advantages are not part of science fiction, and 3D printing technology has achieved some of the above functions. As 4D printing technology continues to mature, it will bring us into a new industrial age where our manufacturing methods and lifestyles will be rewritten by printing.

Section 2 | Three Major Constraints to 4D Printing

Judging by how 3D printing has developed so far, behind its seemingly bright prospects, it still feels that it is short in self-appreciation. The constraints in its device upgrades, such as printing speed and stability of printing effects, has affected how a 3D printing device can effectively play the role of an interconnected media platform in the IoT. Besides, it also fails 3D printing when favored by venture capital.

Similarly, 4D printing, which is inherited from 3D printing and has just embarked on its early stages of development, has been facing a series of objective bottlenecks, such as equipment, materials, and techniques. How can we overcome the constraints to transform 4D printing into a breakthrough? This is not only the thinking of cutting-edge and front-line technologists but also where the opportunities are for the emerging technology entrepreneurs.

So far, the bottlenecks that constrain the development of 4D printing are mainly in the following three aspects.

1. Equipment Constraints: Breakthrough for the popularity of 4D printing technology

In the world of 3D printing, we have to use large-scale machines to print out the large-scale materials used for large-scale projects. These include construction and pipelines, which place high requirements on the printers, including volume, accuracy, reliability, among other factors. This has created a real-life barrier for prompting and popularizing 3D printing technology.

In the face of the lessons learned from 3D printing, 4D printing has taken another path, which is, relying on "smart" materials' to give printed objects the ability of self-transformation. This largely alleviates the constraints of printing devices to technological development. But, "smart" printing also means higher technical requirements for printing devices which will inevitably lead to the high prices of the printing devices. This will in turn result in the fact that 4D printing has to go through a long period of being a "plaything of the rich" before it moves towards technology development and popularity.

Other than the price of small-scale 4D printers, also known as household-level printers, printing accuracy, operational expertise, and other aspects are still in the stages of being fully developed. These factors that currently constraint the rapid development of 4D printing have also left an uncultivated land for entrepreneurs interested in 4D printing. How could we effectively manufacture equipment and enter the market precisely to determine the success probability of entrepreneurs?

2. "Smart" Materials: Pioneers of 4D printing

In addition to the constraints of the printing device itself, 4D printing is facing a more severe problem than 3D printing – the requirement of special printing materials. Compared with 3D printing, the requirements for 4D printing are much higher.

First of all, what 4D printing requires is not ordinary materials, but smart materials with memory functions (Figure 9-1), which are a new type of

Figure 9-1 Memory Smart Material

functional material that can sense external stimuli and then self-transform and self-assemble through judgment.

Secondly, the materials need not only to be 3D printable but also have the sensing function, feedback function, information recognition, and accumulation function, response function, self-transformation capability, self-assembly capability, self-diagnosis capability, self-repair ability, and super adaptability, as well as capabilities of rapid-response transformation and assembly.

Currently, 4D printed objects are still only able to sense the stimulation of water. In other words, the current trigger medium is limited to water only. With the advancement of science and technology, one day in the future, we will be able to realize the materials' self-transformation by triggering the stimuli of light, sound, heat, water, gas, temperature, and any other arbitrary medium. That is to say, while 4D printing is facing infinite challenges on the long road to success, it is also facing endless possibilities and opportunities. Just as there is a risk of failure and a probability of success for each particular experimented material and test, the selection of 4D printing materials and the attention and investment to their applications will become an important direction and development path for entrepreneurs.

3. The R&D of Software: Advancing 4D printing

The reason why 4D printed objects can self-transform and self-assemble at a specific time through the triggering of the medium, other than the printing material itself and printing technology, is the design of the software used to print them. This means, while completing the design for the printing materials, the transformable elements will be directly implanted into the material. A simple but not very accurate understanding is that 3D printing means designing products, while 4D printing is embedding the printer in the printing model. Therefore, for 4D printing, design software is critical and is one of the links in the development that needs to be advanced. How can we de-professionalize design software so that users can operate and apply it without prior knowledge? This is one of the indispensable factors for 4D printing's popularity.

With the gathering of capital and talent, much how they have once advanced 3D printing, the problems that constraint 4D printing will soon be solved, too. This will surely lead to a new round of opportunities for the manufacturing of 4D printing devices and the development of application software. More importantly, it will promote and drive the development of new materials.

Section 3 | Six Major Trends in 4D Printing

On top of 3D printing, a self-changeable dimension is added to it for 4D printing. This brought profound changes and influence on the development of both industries and human society. When such science fiction technology enters our life and becomes a reality, it will affect and change the current business forms and will trigger new development trends.

1. 4D printing will transform industries

Whether it is Industry 4.0 or 3D printing, what they have solved is only the shift from the past mass production to customization, but not the products' self-assembly and self-transformation needs. Although 3D printing technology has been a more direct help for the design of complex structures, including making some structural components lighter and stronger, it is still unable to self-drive and change in response to the external environments.

Despite the fact that 3D printing technology has been applied in industries such as aerospace and construction, it is still difficult for it to meet the changes in different environments. For example, a 4D printed aircraft can decompose itself while facing specific environmental changes to provide the best protection to its passengers. The same could be done in aerospace, whereby we wouldn't have to carry the huge assembled equipment into space, but merely the materials and a 4D printer, or the printed materials, and under the trigger of a specific medium, the space station, space suit or other operating equipment could automatically assemble and form for us.

For everyday industries, 4D printing technology will revolutionize the current manufacturing forms and business forms.

2. 4D printing could help to overcome cancer

The feasibility of using 3D printing to print organs and limbs and effectively match them to patients has been proven in the medical field. The same is also possible for some relatively bony limbs. However, when it comes to the printing of some soft tissues and organs, the performance of 3D printing technology is not as amazing as that of 4D printing technology.

The most disruptive aspect of 4D printing technology is not the printing of organs or limbs, but the printing of cells. By printing drugs in the form of cells that are nano-scale and similar to human cells, and then injecting them into the human body, and setting the trigger media as the corresponding cancer cell virus, 4D printed drug cells could patrol the human body. They will be able to release the drugs contained point-to-point to kill the encountered cancer cells. The hope of humankind to overcome cancer perhaps is pinned on 4D printing technology.

For unknown viruses that humans are exposed to, with the help of 4D printing technology, the same method can be adopted, which will be injecting a harmless vaccine into the human body.

3. The popularity of private factories

Before 3D printing technology, personal customization was almost impossible, and all our consumption could only depend on industrial mass production. For manufacturers, to reduce costs and improve enterprise benefits, they needed to integrate supply chains for different components and adopt large-scale industrialized, standardized, and automated production. Therefore, it was not easy to meet the demand for personalization.

But when it comes to 3D printing, it is easy to fulfill the demand of personalization or even personal customization based on users' preferences. However, what 4D printing can change is not limited to personal customization, more importantly, it could allow each user to own a "factory" that can define and produce the items they want. Imagine you want to buy a refrigerator. You don't have to buy an entire huge refrigerator for your home, but you could buy the core components, or you can even define and print the components out on a 4D printer, and then the printed components will automatically assemble into a refrigerator for you.

This is very different from personal customization, whereby we can express our ideas, build our lives, and even define our innovative "factory" at home with the help of 4D printing technology. What's even more amazing is that when we need more space for the freezer compartment and less space for the cooler compartment, the refrigerator could automatically trigger to assemble and adjust the space of the compartments accordingly.

4. The speed of Innovation will be further boosted.

One problem faced by traditional industrial manufacturing is the long time cycle from product establishment to development and then to market launch, which usually takes at least 6 to 24 months. This is because it involves the stability of batch manufacturing, as well as the manufacturing and service life of molds. All these issues constrain the time cycle of product innovation.

But with the emergence of 4D printing technology, not only our ideas will be printed quickly, but we can also embed a variety of possible corrections into the models. After the object is printed out, we can trigger it to self-transform, self-refine, and self-correct according to our needs, which will indeed be a meaningful thing for people who are pursuing innovation, personalization, and differentiation.

The rapid realization of personal ideas will further reduce the time cycle of innovation, and much the same will be done for the traditional industrialized enterprises. Perhaps shortly, what traditional factories sell will not be the products, but rather years of technologies and services. Users could purchase technical and manufacturing experiences to assist them in the innovation of products based on their ideas.

5. The appearance of new business models

With the arrival, maturity, and popularization of 4D printing technology, we may not sell the completed products in the future supermarkets but sell some key components instead. So, users could buy the needed components and then connect them with the printed ones to realize innovation.

The business models brought about will cause significant changes. On the one hand, the current sales model of supermarkets will change from selling completed products to providing 4D printing customization services and technical support. On the other hand, people engaged in creative work will obtain commercial value by providing consumers with creative designs and 4D printing services. Besides, small chain supermarkets mainly selling a variety of components will emerge.

More than bringing new business models, for IKEA that focuses on selling creative designs and leaves the assembling job to customers, 4D printing will completely change its business model. 4D printed products will not need any manual assembly.

6. The issue of intellectual property will be magnified
4D printing and 3D printing are both facing one common controversy, that is, copying the copyrighted product designs. As such, this issue of intellectual property will be magnified. When a user sees a creative work they like, they could take a picture of it, generate the corresponding 3D model by using software on the computer and then input the model to a 4D printer for printing. This is when the protection of the intellectual property will become extremely tough.

In particular, 4D printed products could even self-assemble and self-change, which increases the difficulty of obtaining evidence. Intellectual property protection is a new issue faced by the new technology era.

Section 4 | Prospects and Challenges of 4D Printing

Compared with 3D printing, which can be described as a young man in his thirties, 4D printing is only a baby. So, it is normal that 3D printing is being sought after. From the observation of foreign countries, the research and trial of 3D printing (also known as additive manufacturing) technology has been around for almost 30 years. It's only in recent years that with the help of capital and accumulated industrial technologies, 3D printing technology has been gradually applied in large-scale applications and extended from industrial to life applications.

From the current development and application of 3D printing technologies, manufacturing industries, including the military and aerospace industries, are gradually showing the revolutionary significance of 3D printing. In fields with special technical requirements like military and aerospace industries, the structure of some components is exceptionally complicated (Figure 9-2) due to the limitations of technical principles and requirements, and so the complexities are either difficult to process under the traditional manufacturing technology or the efficiency of processing and manufacturing is extremely low. It could also be that the manufacturing cost is high and the manufacturing time cycle is long.

Figure 9-2 Aerospace Components

3D printing together with the technical principle of overlay has thus made some complex structures relatively simple and it has significantly improved the processing efficiency. Therefore, 3D printing has been favored and valued by the military and aerospace industries. Before the end of this wave, the U.S., who is the leader in industrial innovation, has entered a more in-depth exploration — the research of 4D printing technology, or the often-called four-dimensional printing technology. This technology has obvious differences from 3D printing technology, and its application fields are broader and more advantageous. It can be regarded as a revolutionary manufacturing technology.

According to the relevant report in the U.S. on May 12, 2014, the Atlantic Council of the United States completed a research report entitled *The Next Wave: 4D Printing – Programming the Material World*. The report was based on another research report completed by the Atlantic Council 3 years ago, *Could 3D Printing Change the World?* discussed this new disruptive technology. It was also the first time that the scientific community formally discussed and deliberated the issues of 4D technology and defined 4D printing technology as the next wave in tech, showing its importance.

10

THE FUTURE (PART 2)

The maturity and popularization of 4D printing will bring not only innovations to human manufacturing technology but also transformations to the future social ecology. The process of making dreams come true will comprehensively stimulate the innovations and creative ability of all people, including the field of children's education and development. It will be a brain-opening spectacle.

With 4D printing, there will be nothing you can't do, only what you can't imagine. The future world will be transformed beyond human imagination.

Section 1 | Humans' Creative Ability Stimulated by 4D Printing

3D printing makes it possible for Makers, designers, and creative people to express and actualize their ideas. It also makes these ideas commercially valuable. It is a consensus before 4D printing technology that to realize the dream of becoming a Maker may have to adopt the development of 3D printing technology.

Experts believe that desktop 3D printers will have an impact on the real manufacturing industry or its industry chains, especially for household goods. This point of view is based on the convenience of 3D printing and the technical principles that can rapidly actualize users' ideas. However, 4D printing technology will not only disrupt the manufacturing industry or its industry chains but also stimulate humans' potential creative ability on a much deeper level.

The four dimensions of 4D printing are beyond our recognition of three dimensions, so 4D printing itself is a kind of creative technology. Its critical difference from 3D printing technology is that under the action of external triggering medium, 4D printing will self-transform and self-assemble, which will effectively stimulate humans' creation ability and bring along the following changes.

1. Releasing human labor

This can be understood on two levels. The first is that the inspiration of most designers and Makers is currently actualized by handcraft. The problem brought about by this method is the limitation of handcraft itself, which means the handcraft masters have to experience years of honing and settling into their craft, and even one technique may require years of work. Even if the handcrafted technique has been mastered to a very advanced level, it is still almost impossible to produce multiple identical products, and this is the limitation of the designer's creations.

Some people may think that this kind of scarcity will make the product precious. But, for those masters who integrate the aesthetic creations with commercial value, certain basic handcraft techniques take too much time, which is a great limitation to their creation value. 4D printing technology not only gives designers more creative space but also a choice to let the handcraft masters further build up the 4D printed models with their excellent techniques.

The other level is the difficulty of second replication. With the integration of 4D printing technology and 3D scanning technology, the second replication can be realized by fusing it with 4D printing technology in the early stage of creation. Alternatively, for handcrafted products, 4D printing technology can be used to produce the second copy.

2. Bringing dreams into reality

Leaving aside 4D printing technology, it is difficult for individuals to express their ideas in the physical world and make those ideas become usable commodities without the use of 3D printing technology. Perhaps we can use traditional machining techniques such as turning, milling, planing and grinding, or machining centers to process the three-dimensional physical objects, but this is still difficult for some complex components.

4D printing technology fundamentally solves the problems of forming arbitrary structures and shapes, allowing self-assembly capabilities of some products to assemble and save on labor. More importantly, the problem of transportation and storage is solved as well. Take a table as an example, its transportation usually happens two ways using the traditional methods. One is to disassemble the table into small unit parts, transport them to the designated location, and then manually assemble the unit parts into a table. The other is for the seller to assemble the table first, and then transport the whole table to the designated location. The disadvantage of the second way is that the table occupies a huge space in the transportation process. However, this disadvantage doesn't exist in 4D printing technology because we can print the table layer by layer into a plank first. Then, after transporting the plank to the designated location, it will self-assemble and self-transform into a table by triggering its medium.

With 4D printing technology, the *Transformers,* the magical paintbrush of Ma Liang, and the Monkey King's Jingu Bang in our minds will no longer be stories. Designers, Makers, or the general public, as long as we have creative ideas, 4D printing will make them a reality.

Furthermore, with the maturity and popularization of 4D printing technology, it will further stimulate the innovation and creativity abilities of all people, including their application in the field of children's education and development. It will bring us a brain-opening spectacle. Perhaps in the future, we only need one book, which will become a corresponding textbook in different classes. Humans' creativity will be influenced by the application of 4D printing technology, and humans' potential will be continuously stimulated as well.

Section 2 | New Career Paths for Industrial Designers

Design plays a key role in a country's economic transformation and upgrade, and industrial design plays a significant role in promoting the current upgrade of China's manufacturing industry. With the global economic slowdown in recent years and the rising cost of China's domestic manufacturing, many enterprises relying on low-cost competition have found themselves in a dilemma. How could they increase the added value of products through design? (Figure 10-1) The creation of the competitiveness of Chinese manufacturing seems to be a concern for most enterprises in China.

Moreover, impacted by the internet, the traditional manufacturing industry, and the consumers' consumption philosophy have undergone profound changes. Concepts of personalized consumption and personal customization are on the rise. The most obvious of the changes caused by these concepts is the formation and emergence of groups represented by the Makers. They undoubtedly impact the traditional manufacturing industry, which is concentrated on the advantages of standardization and mass manufacturing. The continuous maturity of 3D printing technology provides the foundation for the realization of such creativity-led personally customizable consumption products (Figure 10-2).

Figure 10-1 Creative Design

Figure 10-2 3D Printed Cups

For practitioners engaged in creative design, especially industrial designers focusing on manufacturing products, 3D printing technology is now used to produce and manufacture the ideas in their minds into physical products. This has become a new opportunity and a trend to materialize ideas, and then realize their commercial value via e-commerce platforms.

But in the era of 4D printing, everything will be subverted. Designers will become ubiquitous "make-up artists" and "beauticians" of products. This means that soon, designers will be able to open up design studios with their style and characteristics on the cloud service platform, and any consumers could be their customers. A consumer either could express their creative ideas via software tools on the computer or they could choose a designer who fits their creative style and convey the ideas to the designer through text, picture, voice message, or any other form. Then, the designer will come out with the design in response to the consumer's request.

Alternatively, a consumer could post their ideas indicating a remuneration they are willing to pay on the platforms that provide creative design services, and any interested designer could accept the job and provide the corresponding service. For Makers, after the creativity based on the technology will be realized, Makers will dress up the products, allowing them to walk into consumers' everyday life from the lab. The biggest difference between 4D printing and 3D printing is that there will be more possibilities and challenges to designers, such as the beauty of the product under different forms, or the best way to change products into different shapes. All these challenges will test the designers' wisdom and will also bring some changes to the design discipline.

Section 3 | The Popularity of Xiaomi

In China, the once-popular Xiaomi smartphone known by many people has become a classic case many visiting professors discuss. "Xiaomi gathers a group of' reliable people who use the Internet to make smartphones."

According to Lei Jun, the founder of Xiaomi, 600,000 enthusiasts who call themselves "Mi Fans" were involved in the development of the smartphone system. From this, it makes Xiaomi smartphone was attached to a large number of enthusiasts, and more and more innovations that meet the habits of Chinese people's use have been created by "Mi Fans" and put into use by the Xiaomi smartphone. The open design of the Xiaomi smartphone has firmly attracted 700,000 enthusiasts before its launch.

The same successful case of open design has also been proved by Nokia's 3D printing community project. On January 18, 2013, Nokia's 3D printing community released a "3D printing development kit" to help people design and make their own Lumia 820 phone case. As introduced by John Kneeland, a Nokia Community, and Developer Marketing Manager, the Lumia 820 smartphone has a detachable case, which has anti-shock and anti-dust protection. Customers can replace the case with different colors, and for the advanced version of Lumia 920 and the intermediate version of 820, wireless charging capability could even be added to the case. These special features give most users of Lumia 820 great options. Other than that, Nokia also released 3D templates, instruction manuals, recommended materials, and the best practices, so that anyone familiar with 3D printing could customize a Lumia 820 phone case for themselves. Only a few days after the announcement was posted to the Nokia community, a Nokia 3D printing challenge for Lumia 820 was held to encourage people to share design ideas of phone cases that could be 3D printed and be used as replacements. Six days after the release of the 3D printing development kit, the 3D printed Nokia Lumia 820 phone cases, containing some functional buttons, were displayed online.

The same creativity will be fully exploited in 4D printing. With the frequent evolution of digital models, product design, and production plans based on the storage and transmission of digital models will be continuously improved and upgraded. This will be done by the second innovation wave from different printing enthusiasts, and the printed products will be evaluated and tested during the printing process. Afterward, the enthusiasts will be able to upload the improved design plans and digital models, which not only will allow others to benefit from the design improvements but also give designers and other printing enthusiasts a new direction of thought.

With the development of computers and the Internet, the increasingly barrier-free spread of cyberspace will give birth to "open design" in the 4D printing era to root, breed, and further spread its ideas. As more and more people will use CAD software, 3D and 4D printing technology, and more people modify and implant their ideas to modify and innovate the designed digital models through the Internet – they will gradually turn the future product design process from exclusive to an open one.

Section 4 | DDM Makes Customization Popular

Mass production is the best way to reduce production costs in a traditional manufacturing environment because the development cost of abrasive tools could be diluted with the increasing production. At the same time, the cost could become very large if the developed abrasive tools are only used to produce few products, which is why the customized products are often expensive.

The application and promotion of "Direct Digital Manufacturing" (DDM) will undoubtedly bring customized products into the homes of the common people. Everyone will be able to design and implant personalized elements into their products, and the printer will only perform the printing job according to the data model received. It's will be just how we print our files on a 2D printer, it does not matter if the content of the files is numbers, Chinese characters, or colorful pictures.

Those who have visited the Cubify.com website are probably familiar with the case of customizing the Apple mobile phone model, which can produce a 3D finished product with a raised back based on the digital photo provided by the customer. Similarly, ThatsMyFace.com invites visitors to upload photos of their front and side of the face, and then generates a colorful 3D head model, which can be built onto a plastic doll or Lego model. Moreover, with the continuous advances and maturity of 4D printing technology, the printed head model will be able to show signs of growth or aging in the natural environment over time. Such customized and changeable products will enable people to constantly feel the rapid changes of the world, and it will no longer be hard for us to easily own a small superhero figurine prototyped after ourselves.

There are already too many similar cases. Personalized custom design and low-cost product manufacturing will terminate the traditional mass production. Instead, it will replace it with the low-cost production of disposable products or small batches of components because special abrasive tools will no longer be required to start the production. For example, the producer of the 2012 James Bond film, *Skyfall*, needed three Aston Martin DB5 model cars that were one-third the size of the actual car, and they printed them out on a 3D printer model Voxeljet VX4000. These model cars were 3D printed into 18 parts and then assembled and spray-painted before being turned into very realistic and expensive-looking replicas. Likewise, a custom motorcycle manufacturer called Klock Werks Kustom Cycles, to manufacture a motorcycle in five days, used engineers that designed the dashboard, fork tube cover, headlight baffle, pedals, pedal bottoms, and wheel gasket covers by using a 3D model software called SolidWorks. Finally, they printed out the model using a 3D printer. The speed of the finished motorcycle even set the land speed record of the American Motorcycle Association. In the foreseeable future, the parts of this motorcycle will even be able to achieve self-transformation, self-extension, and seamless splice under the action of a certain medium with the application of 4D printing technology. This will further simplify the printing process and improve the printing efficiency.

Many successful cases have proven to us that by 4D printing technology relying on 3D printers, it can produce various personalized products quickly and conveniently. Plus, the products could transform and upgrade within the time dimension. The low cost and techniques of the products are unachievable by traditional manufacturing methods. For either enterprises or individuals, those who want to produce one or a small batch of the latest trendy products, 4D printing will be an unparalleled choice.

Section 5 | The "Magic" Effect of 4D Printing

4D printing, which follows the pace of 3D printing, will lead us to new dreams and make science fiction a reality. While Captain Jean-Luc Picard was sitting in the preparation room of the Starship Enterprise and wanted to have a hot drink, he just needed to verbalize it in words, and the "replicator" on the starship could collect all the necessary atoms, including the atoms to make a cup, and then make him a cup of hot tea. The common use of the "replicator" to Captain Picard is similar to our use of the microwave today. The microwave oven uses wireless radio waves to activate atoms and generate heat, which was an exciting event in the 1950s. While the unexplained magical technology of the "replicator" collects and aggregates atoms to form food and drinks.

Star Trek: The Next Generation is a science fiction film. However, Picard's "replicator" is not completely impossible to turn into reality. When you add a little poetic imagination to the current industrial 3D printers, you will see the "replicator" in *Star Trek.* In a basin of stationary liquid resin, a laser begins to track the shape of the resin-like lightning to produce various shapes, and then as if there was magic that made different kinds of food, it suddenly appeared. With the advent and popularization of 4D printing, these foods and items could even change according to our expectations and needs under different time dimensions and actions of media.

Let's put aside the poetic imagination as we are still far away from the self-aggregating atoms, or at least a considerable distance from their effective use. However, through 3D printers and 4D printing technology, we can realize the personalized custom production we need. As long as we can present the ideas on the computer and shape them with a machine, just by simply pressing a button, the objects will appear like magic. Perhaps this is the saying, "Any sufficiently advanced technology can be called magic."

1. Lower costs: No more expensive customization fees

When 3D printing was first introduced, some scholars used the analogy of a platypus.

When the platypus was first spotted, explorers thought it was a hoax, thinking it was a joker who stuck a duck's beak, webbed feet, and a kangaroo's pouch on a furry animal.

Similarly, 3D printing is the "platypus" of the manufacturing industry, combining precise digital technology, factory repeatability, and the craftsmen's design freedom.

Just like the machines in the factory, 3D printers are also automated. The digital design file will concisely receive the instructions for producing a specific product, and then guide each step of the 3D printer. This process can be saved or sent anywhere via email. Just as craftsmen can produce multiple products, 3D printers have multiple uses. A 3D printer can produce a wide variety of products without a large upfront investment.

Following in the footsteps of 3D printing, 4D printing has the same effect. We can 4D print 1,000 different products and 4D print 1,000 identical products, and the costs will be the same. The cost of customization virtually disappears because the development and manufacture of abrasive tools will no longer be needed.

In other words, the application and promotion of 4D printing have removed the shackles of any conditions for the manufacturing industry's personalized and customized production. Even though traditional mass production has a high-efficiency advantage that can increase enterprise profits and reduce retailing prices, the traditional economies of scale harm product diversification and customization. Conversely, craftsmen can easily produce many diversified and customized products, but the output will be extremely restricted. With the arrival of the era of 4D printing and digital manufacturing, we will be able to choose between mass production and customization without having to pay expensive handcrafting costs.

Along with the recognition of 3D printing, people's acceptance of 4D printing has rapidly improved. For enterprises whose business model is to sell a small number of unique, customized, or changeable products with high marginal revenue, the development of 3D printing and the application of 4D printing will undoubtedly bring revolutionary leaps forward. This will be of epoch-making significance.

2. More efficiency: The manufacturing of the product's prototype can be broken down

As our world accelerates, enterprises are increasingly eager to shorten the time it takes to go from the design stage to product delivery. The production cycle is a key measure of enterprise efficiency, which means that the shorter the time between design and completion of the final product, the better.

4D printers will minimize product delivery time by enabling designers and engineers to efficiently produce prototypes at low cost on-site. The prototype is the first draft of the product, which helps designers, engineers, marketing teams, and manufacturers to conduct multiple checks to see how the finished product will look. Previously, we provided rapid prototyping services to automotive manufacturers through 3D printing. By 3D printing out the design concepts and presenting them to the project team, the automotive manufacturers saved time and even saved consumers' time.

In comparison, the past prototyping was a lengthy and costly process, and it was risky for manufacturers to take shortcuts or to be overly confident in the fact that product design could be realized. Even if the designs were beautifully made, manufacturers could still find many other problems after they were produced because it was difficult to visualize the designs in CAD models.

For example, if you own a car, you need to repair and refuel it. But we cannot simply make a hole in the engine to refuel the car, and human hands wouldn't fit like that, either. Therefore, we can imagine that 3D printed prototypes will soon replace those handmade with foam or clay. Many enterprises have skipped foam and clay prototypes and directly adopted 3D printing prototypes. After 4D printing is implemented, everything will be further sublimated.

Through 4D printing, we can even perfect the design and further shorten the delivery time based on 3D printing. The 4D printed prototypes can change at a specific time and under the action of external media so that the printed samples can more effectively and accurately fit the product design and application requirements. Instead of us trying to achieve perfection by printing the product again and again. In this way, the time cost will be shortened and production efficiency will be enhanced.

3. Better quality: Manufacturing the best-customized components

The printing of customized terminal components is one of the fastest-growing applications of 3D printing, and the customized component will be one of the leading products of 4D printing, too. So, the customized components will no longer be prototypes but real products. For people who have a 3D printer at home, they could log on to the community forum to exchange ideas and design documents of standard replaceable components, such as door handles and shower curtain rings, which could also be rapidly communicated, exchanged, and applied in 4D printing. The 4D printed components include door handles, gears, antiques, or discontinued components that are hand-made and extremely expensive.

As production of customized components is not profitable in the economies of scale, small-scale and technical 3D printing service providers and 4D printing operators will start looking for new business opportunities. Automobile and motorcycle manufacturers (and even Mars rover manufacturers) are using 3D and 4D printed custom components to produce concept cars or machines. After all, a multi-million dollar vehicle is too expensive for test drives.

4D printed terminal components are widely used in the medical industry and the dental industry because the products of these industries require closer and more accurate integration with the human body. The suitable braces and tooth crowns for patients used to be customized, but now, they are increasingly 3D printed. In the future, the transformation capability of 4D printing will make it an even better choice. Likewise, by scanning a patient's ear canal or residual limb, the scanned data could be used to print out corresponding hearing aids and prostheses to maximally fulfill the needs of the patient under the action of time dimension and media.

4. Cheaper: No wasteful consumption and low-cost production

4D printing will minimize consumers' passive, wasteful behaviors. This is because 4D printed products could transform under the action of a certain medium along with the change in time and space dimensions. This will satisfy consumers' needs to the greatest extent. To give a simple example, the growth of newborns is "changing with each passing day," and whether it is clothes or shoes, there will always be the "embarrassment" of them being too small after not wearing the items too long. But, 4D printed children's clothes or shoes will have the magic of being able to "grow up" with the children.

The consumption of 4D printed products can reduce unnecessary waste for consumers, and the production of 4D printed products can also reduce the cost in the product development process for manufacturers. Some 4D printed prototypes are used to demonstrate design concepts, whilst others are used to test stages of products' life cycle to figure out how to achieve mass production of the products' components. The "test and adjust" ability of 4D printed components can not only maximize the variations of the products in different time dimensions and under different media conditions but also minimize the inherent defects of the complex-designed products. This has also been fully proven in 3D printing.

The prototype for testing and adjusting can be the non-assembled components used by engineers in a simulated production process. Earlier, when Microsoft announced the top-secret product concept (the hybrid tablet and laptop the Windows Surface), the world was shocked. The media wanted to know how Microsoft had developed this product under confidentiality. Usually, when a company first announces the launch of a new, cutting-edge technological product, mysterious photos will leak out of its manufacturing factory. How Microsoft has done this is that the product's prototype was 3D printed in the hardware department located on a university campus, ensuring that product development was being carried out secretly.

The other purpose of testing and adjusting was to make sure that the machines in the factory could produce the exact product according to the design concept. In the engineering product design course, students spend weeks choosing between the great design ideas and the realities of the factory. Although those thick textbooks detail which design concepts are and are not feasible on the production line, it is hard for common molding machines or cutting molding machines to manufacture hollow objects, interlocking parts, or products with complex internal structures. Not all production challenges can be avoided under the guidance of the textbooks.

If an enterprise finds out too late that the components of the new product do not fit together, the initial investment will be wasted. The 4D printed product's prototype for testing and adjusting can help designers of mobile phone products assemble tiny hardware, such as hearing aids, car steering wheels, razors, combs, and smartphones, which must be comfortable to the touch or fit snugly on the user. Even though design software and computer simulation are becoming more and more advanced, even the best designs may not always be produced exactly as planned.

Section 6 | The Future World is Beyond Imagination

The most direct understanding of 4D printing is that the time dimension is added based on 3D printing, and then time and corresponding models to be transformed into the printing model is embedded with the help of software.

This allows the transformable printing materials to automatically transform and assemble into the predetermined shape under the action of a certain medium at a specific time.

The biggest change is this process does not need to resort to complex electromechanical principles, or complex computer programs and electromechanical equipment. Instead, it prints a kind of self-transformable material, which is embedded with the assembling and transformable design capabilities through software. Then, at a specific point in time, the printed product can fold, assemble, and transform automatically into the shape conformed to the intention of the user's initial design. From this perspective, we can perhaps make a distinction between 3D printing and 4D printing, and the distinction is that the logic of 3D printing is to build the printing model and then print out the final product. While the logic of 4D printing is to embed product designs and time into the transformable smart material through a 3D printer. So, it can be said that 3D printing is a static printing process, whereas 4D printing is a dynamic shaping process. The key to 4D printing is its ability to self-transform over time.

The key to 4D printing is not the innovation of printer technology, but the innovation of printing materials. Currently, printing materials are mainly based on memory alloy material. The emergence of 4D printing technology will inevitably push the development of the field in new material technology. The recent research results of 4D printing technology, which was the first new technological achievement, was collaborated by MIT and Stratasys' Education & Research & Development. It was a revolutionary new technology that allowed materials to self-transform and self-assemble without the help of external forces. This technology in China is currently non-existent.

According to the founder of 4D printing technology, Tibbits and his team, the current 4D printing is still in its infancy. It can only print strand-like objects to realize the changes from one dimension to two dimensions and from one dimension to three dimensions. Moving forward, they will design and print out sheet-like objects and then three-dimensional objects to realize changes from two dimensions to three-dimensions and three-dimensions to more complex structures. Hence, the *Transformers* that exist in science fiction films will come out of the screen and enter our life.

It is conceivable that in the era of 4D printing, products will become more intelligent and humane. Space stations and satellites will be able to realize self-assembly and self-repair, projects in dangerous areas will no longer require human participation, everything such as bridges, dams, highways, and houses will build on their own according to their design. People will only need to sit at the computer, designing products that suit their ideas and needs, and then simply click "print," and then it's all going to be done.

Section 7 | Opportunities and Challenges of "World Factory"

"World Factory" is a specific term for the world's industrial powers, especially the manufacturing powers. In the history of the world economy, the three countries explicitly known as the "world factories" are the United Kingdom, the United States, and Japan, in that order.

In 2009, China replaced Japan as the world's second-largest trading country after the U.S. and successfully surpassed Germany to become the world's largest exporter. At the same time, China's manufacturing industry accounted for 15.6% of the global manufacturing value, becoming the world's second-largest industrial manufacturing country after the U.S. For a time, China became a globally recognized as a "world factory."

However, times have changed. Nowadays, the cost of hiring a worker in China is currently 1.5 times higher than in Thailand, 2.5 times than in the Philippines and 3.5 times than in Indonesia. China, as the "world factory," has the advantage of cheap labor costs that have gone up, and Southeast Asian countries are ambitious to become the "substitutes." Meanwhile, developed countries such as the U.S. are also advocating "re-industrialization" and "revitalizing manufacturing" in a high-profile way. After all, under the attempt of "de-industrialization," life in these countries is not easy. The price of labor in European and American countries has not risen very much. Under the influence of the economic crisis and with the employment depression, labor prices have even decreased. According to the data reported by *Time* magazine in April 2012, the continued economic recession has brought the number of unemployed young people aged 15 to 24, to 75 million worldwide. In most western European countries, excessive labor protection has made it difficult for young people to find suitable jobs. If this is the case, can people still get high wages? Likewise, given such circumstances, if wages are equal, why should the manufacturing be deployed overseas to create employment opportunities for other countries?

In the case of the U.S., re-industrialization, and revitalization of the manufacturing industry objectively requires the country to adopt policies that can promote the development of the domestic manufacturing industry and reduce the cost advantage of foreign competitors. By itself, the manufacturing industry will be more attractive. Presently, the measures to reduce the cost of the manufacturing industry have already been taken and the U.S. has also taken measures to include tax incentives for companies that relocate manufacturing overseas. Meanwhile, punitive taxes are being charged to those who do not. Among the factors that ensure the U.S. enterprises are committed to designing and producing technologically advanced products that are needed by customers around the world are the reduced taxes and the making of temporary tax reduction measures permanent for the country's manufacturing industry to improve the manufacturing industry's ability to attract capital and investment. On top, there is reforming of the medical insurance and reducing the cost of medical insurance, reducing the cost of regulatory and judicial litigation, implementing energy-saving programs to reduce energy costs, encouraging innovation investments, and finally, promoting the diffusion of technologies and reducing the risk of developing new technologies.

Concurrently, China's manufacturing industry, which has been going smoothly and has been invincible over the past 30 years, has encountered a sandwich competition between high-end technology and low-end costs. When the traditional advantage of cheap labor ceases to exist, what other advantages does China have to maintain its prosperous "world factory" status?

The emergence of 3D printing and the arrival of 4D printing mean there is hope for both European and American countries who need "re-industrialization" and China who wants to maintain its status as a "world factory."

The emergence of 3D technology will cause a rapid reduction in social manufacturing demand for industrial workers. Manpower will no longer be a scarce resource, and traditional factories will no longer exist. As long as there is a printer, production could be done anywhere. As a result, the added value of intellectual costs will further increase in the proportion of social manufacturing. Therefore, the U.S. government is facing the emergence of manufacturing depressions and the employment decline that is particularly interested in the new technological revolution represented by 3D and 4D printing.

Obama once proposed to promote a National Network for Manufacturing Innovation (NNMI) of as many as 15 manufacturing innovation institutions across the country. Each of these institutions had a research focus. One of the research focuses was the improvement of the relevant standards, materials, and equipment of 3D printing technology (also called additive manufacturing) to realize the use of digital designs for low-cost and small-batch production.

On August 16, 2012, the U.S. government announced the establishment of a manufacturing innovation research institution jointly funded by the government and a private sector in Ohio. It mainly developed 3D printing technology to boost innovation and growth in the manufacturing industry and regained the leadership of the world industry. As the leader of the global economy, the U.S. places its hopes of regaining the glory of the manufacturing industry through 3D printing, and the strategic significance of 3D printing is self-evident.

On March 24, 2014, Obama visited the "3D printed Canal House" project during the 53-nation leaders' Nuclear Security Summit in the Netherlands. The project was already an attraction in the Netherlands, and it was no longer printing a pure model, but a full-size "model." These houses were initially tall and narrow brick houses built by wealthy Dutch traders along the canals in Amsterdam. But now, high-tech was easily restoring them.

Similarly, the emergence of 3D printing and 4D printing gives hope but also poses a challenge for China.

According to the data in the article *The U.S.'s Strategy for Promoting the Third Industrial Revolution and Its Reference to China,* the power gap between China and the U.S. is mainly reflected in innovation capabilities. In 2012, the U.S. ranked 10th in the world for innovative technologies, while China ranked 34th. In terms of human capital and R&D capabilities, the U.S. ranked 22nd in the world, and China ranked 84th. This gap is the biggest problem faced by China's manufacturing industry.

In 2012, Vivek Wadhwa, Director of the Duke Center for Entrepreneurship and Research Commercialization in the U.S., published an article *titled The Future of Manufacturing is in America, Not China* on the website of *Foreign Affairs* magazine, believing that "Technical advances will soon lead to the same hollowing out of China's manufacturing industry that they have to U.S industry over the past two decades."

According to Vivek Wadhwa's point of view, the future economic model of humankind will be a creator economy with a 3D printing-based production model. There will no longer be large-scale industrialized production.

At this point, if China can seize the opportunity of 3D printing, its manufacturing industry will achieve more remarkable development. Conversely, if the opportunity is missed, it will not be long before the center of the manufacturing industry will return to the U.S., and China will further widen the gap between the Western countries.

In the high-tech field of 3D printing, God has given the Chinese nation a rare opportunity to compete equally and appear on the same stage with the U.S. and Germany. This is because there is no precedent in this field. Both large and small enterprises start at the same starting line. Unlike other fields, Chinese enterprises are trapped by the "patent pool" set up by those large international enterprises.

"3D printing is one of the signs of the third industrial revolution, and it is also the first time in the history of our country that we have the opportunity to participate in the rise of a new industrial revolution." Dai Rong, the Academician of Shanghai Ninth People's Hospital affiliated to Shanghai Jiao Tong University School of Medicine, expressed that the difference of 3D printing technology between China and developed Western countries was not very big. Western countries have developed multi-nozzle printers, and China can also do the same. Western countries have relevant patents, and China has many patents, as well. There are many printing materials that are even superior to those of Western countries.

Since the 1990s, many universities in China have been successively carrying out independent research and development of 3D printing technology.

Tsinghua University has scientific research in some areas such as modern molding theory, Laminated Object Manufacturing (LOM), and Fused Deposition Modeling (FDM). Huazhong University of Science and Technology has advantages in LOM technology and has introduced Horseradish Peroxidase (HRP) series molding machines and molding materials. It has independently developed the 3D printer nozzle, the Stereo Lithography Apparatus (SLA) system, and the corresponding molding materials with a molding accuracy of 0.2mm even before the Jiao Tong University. It also independently developed an eight-head nozzle device, which is expected to be applied in the fields of micro-manufacturing and optoelectronic devices.

Although China and Western countries are on the same starting line of the emerging industrial revolution, it is undeniable that China lacks the environment for the growth of innovative talent, and the entrepreneur's lack the spirit of "persistence, professionalism, and dedication." So, it is a great concern for China to value the focus and investment in 3D and 4D printing technology.

At present, there are no professional institutions engaged in 4D printing in China, and the total number of scientific research units and enterprises engaged in 3D printing technology is only about 20 or 30. The output value of the largest 3D printing enterprise is less than 100 million, and 10 million to 20 million for most of the other enterprises. Moreover, 3D printing technology enterprises still adopt the traditional industrial enterprise operational mode by taking the old path of producing and selling equipment. This has resulted in the low industrialization and acceptance of 3D printing domestically. The capability of R&D opportunities for 3D printing technology in China is still far below that of Western countries.

First, the lack of innovation will restrict the demand for 3D printing in China. Processing manufacturing is still the mainstay of China's manufacturing industry, and product creativity and innovation are not accomplished overnight. There is also a contradiction between large-scale manufacturing and technological innovation. Secondly, China's 3D printing equipment lags far behind the international levels in terms of printing accuracy and equipment reliability, and the core components that restrict the development of 3D printing still rely on imports. Besides, the industry will have to encounter risks such as technological bottlenecks and investment withdrawal while maturing.

So now, China has to grasp opportunities and courageously face its challenges.

China is vigorously promoting the implementation of industrial transformation and upgrade strategies. Industrial transformation and upgrades are essentially an upgrade to the high-value link of the industrial chains, which means that China's industry must shift the advantage of competition from being low-cost driven to becoming innovation-driven. To complete this industrial upgrade, it must be driven and supported by three major innovations. The first is institutional innovation, which is the decisive condition for China to become an industrial power. Second is scientific and technological innovation, which is the key factor for China's industry to surpass the Western countries. The third is education innovation, which will serve as a guarantee for China's industry to have a large number of innovative talents to build industrial power.

This world has been quietly changing. If you find this out one day, it means the world is already different. It's just like investing. In the investment world, while looking back, you will see so many investment opportunities around you, and you will start to regret why you haven't chosen that investment opportunity. But looking forward, it's all abyss. Any business is full of risks, and opportunities are only reserved for those with judgment and foresight.

ENDNOTES

1. Baidu is a popular online search engine in China.
2. The term refers to a group of people who love technology and its practice, and they enjoy sharing technology and exchanging ideas. The community (Makerspace) with Makers as the main body has become the carrier of Makers' culture.
3. This is a fairy tale from China. The story is about a little boy, Ma Liang, who obtains a magic paintbrush by coincidence. No matter what he draws using the paintbrush, it becomes real. The little boy uses this magic paintbrush and his wisdom to help the poor against the oppression of the emperor.
4. Monkey King is the character of a Chinese novel, Journey to the West. Jingu Bang is a magical cudgel wielded by Monkey King.
5. RMB stands for Ren Min Bi, the currency of China.
6. A Transformer is a giant monster from a movie.
7. Jerry Qi is a former associate professor of mechanical engineering at CU-Boulder, and now he is an associate professor of mechanical engineering at Georgia Institute of Technology.
8. Jerry Qi is a former associate professor of mechanical engineering at CU-Boulder, and now he is an associate professor of mechanical engineering at Georgia Institute of Technology.
9. Year-Over-Year (YOY) is a frequently used financial comparison for comparing two or more measurable events on an annualized basis.

ENDNOTES

10. The golden shield and the iron shirt can sustain the thrusts of sharp weapons on one's bare skin.
11. UAV stands for Unmanned Aircraft System.
12. VR stands for virtual reality.
13. The Internet of Things (IoT) refers to the billions of electronics around the world that are now connected to the internet to send data, receive instructions and share data, which doesn't need human-to-human or human-to-computer interaction.
14. The Wind Fire Wheel is one of the well-known magic weapons in Chinese mythology. The sound of wind and fire can be heard when the two wheels rotate. The wheels belong to Nezha who is an ancient Chinese mythical creature and is a legend of the immortals. While Nezha's wearing them, he can freely fly.

INDEX

INDEX

INDEX

ABOUT THE AUTHOR

Dr. Kevin Chen, or Chen Gen, is a renowned science and technology writer and scholar, postdoctoral scholar at Boston University, and an invited course professor at Peking University. He has served as special commentator and columnist for *The People's Daily*, CCTV, China Business Network, SINA, NetEase, and many other media outlets. He has published several monographs involving numerous domains, including finance, science and technology, real estate, medical treatment, and industrial design. He has currently taken up residence in Hong Kong.

CPSIA information can be obtained
at www.ICGtesting.com
Printed in the USA
LVHW090800110521
686847LV00002B/2

9 781487 804886